Digitalization of Power Markets and Systems
Using Energy Informatics

Umit Cali • Murat Kuzlu
Manisa Pipattanasomporn • James Kempf
Linquan Bai

Digitalization of Power Markets and Systems Using Energy Informatics

Springer

Umit Cali
Elektro E/F, E427, Gløshaugen
Norwegian University of Science and
Technology
Trondheim, Norway

Manisa Pipattanasomporn
Smart Grid Research Unit (SGRU)
Chulalongkorn University
Bangkok, Thailand

Linquan Bai
Systems Engineering & Engineering
Management
University of North Carolina at Charlotte
Charlotte, NC, USA

Murat Kuzlu
Electrical Engineering Technology
Old Dominion University
Norfolk, VA, USA

James Kempf
Kempf and Associates
Mountain View, CA, USA

ISBN 978-3-030-83303-9   ISBN 978-3-030-83301-5   (eBook)
https://doi.org/10.1007/978-3-030-83301-5

This Springer imprint is published by the registered company Springer Nature Switzerland AG
The registered company address is: Gewerbestrasse 11, 6330 Cham, Switzerland

Preface

If you were to ask someone what was the most important engineering achievement of the twentieth century, chances are that they would say "the manned moon landing" or "the taming of the atom" or "the invention of the automobile" or perhaps even "the invention of personal computers and the Internet." Practically no one would say "the development of the electrical grid" because the use of electricity is so fundamentally interwoven with our lives that we rarely think about what life would have been like without it. Yet arguably, the grid has had a larger impact on the quality of human life for more people in the twentieth century than practically any other engineering achievement. At the beginning of the century, electricity was expensive and confined to small areas in cities. By the end, practically everyone in developed countries had access to a reasonably priced, reasonably reliable 24×7 electricity supply. The task by the end of the twentieth century then seemed to be to extend this engineering miracle to the developing countries.

Yet the primary energy sources used to power the grid were originally based almost entirely on fossil fuels, and the carbon dioxide emitted from burning them was rapidly accumulating in the atmosphere as access to and demand for grid-supplied electricity grew. Carbon dioxide is a greenhouse gas, and by the beginning of the twenty-first century, scientific studies established that if burning of fossil fuels did not stop – and the sooner the better – the planet would soon become uninhabitable due to global warming. Nuclear power, which emits no carbon dioxide, seemed the logical choice until rapidly escalating development costs and the disasters of Harrisburg, Chernobyl, and Fukushima revealed that nuclear power in its current form is too expensive and too dangerous to scale up to the level necessary to replace fossil fuels.

As the reality of the climate catastrophe became more evident in the early decades of the twenty-first century, some countries began establishing policies to subsidize the deployment of renewable energy as the primary energy source for the grid. Renewable energy comes from environmental sources like wind and solar, but also from water and geothermal. Technologies for utilizing these sources had been available for many years and, in limited cases (primarily hydroelectricity and geothermal generation), have been deployed since the latter part of the twentieth century and even earlier. But in most cases, wind and solar were too expensive to compete effectively with fossil fuels. Because renewable energy sources are variable, storage technologies such as lithium batteries and others are necessary to store energy when the renewable resource is abundant and release it when it is scarce. By the beginning of the third decade of the twenty-first century, renewables had ridden the cost curve down to where solar PV, even without subsidies, became the cheapest form of electricity on the planet.

Due to the urgency of reducing carbon emissions contributing to the global climate crisis, the reign of fossil fuels as the predominant supplier of energy is rapidly coming to an end as the energy sector evolves toward a greener vision, known as the "Green Shift." The integration of digital technology from the information and communication technology (ICT) sectors into the Green Shift has triggered an additional transformation wave by adding digitalization into the picture, resulting in the "Digital Green Shift" (DGS). The DGS is a framework that aims to tackle the challenges and interrelations associated with higher penetration of renewable energy resources but at the same time making sure that the latest digitalization technologies such as artificial intelligence, blockchain technology, and advanced ICT are utilized in an orchestrated way to provide sustainable and affordable solutions.

This book is designed for students entering a power system engineering university curriculum, to give them an introduction to the digital technologies that are revolutionizing the twenty-first century grid. In addition, power systems professionals without a background in digital technologies can use this book to jumpstart their career development in the new technologies. This book also is written to serve as guidance for the engineers or other professionals who would like to gain the foundations of the digitalization of power markets and systems. It explains the basics of the smart grid communication protocols, smart grid cybersecurity, consumer use of digital technology in power systems (also known as the "Energy Internet of Things" or Energy IoT), uses of blockchain, artificial intelligence, machine learning in power systems, and optimization in power markets. Students and mid-career professionals looking for an introduction to smart grid need to understand how these topics fit together with traditional power systems topics in order to intelligently participate and guide the ongoing grid modernization. This book will start them on that journey and act as a guidebook through the complexities of the digital technologies transforming the grid.

We hope you are as excited as we are by the grid modernization revolution toward DGS, and we hope this book helps you get started on your journey.

Elektro E/F, E427, Gløshaugen   Umit Cali
Norwegian University of Science and Tech
Trondheim, Norway

Electrical Engineering Technology   Murat Kuzlu
Old Dominion University
Norfolk, VA, USA

Smart Grid Research Unit (SGRU)   Manisa Pipattanasomporn
Chulalongkorn University
Bangkok, Thailand

Kempf and Associates   James Kempf
Mountain View, CA, USA

Systems Engineering & Engineering Management   Linquan Bai
University of North Carolina at Charlotte
Charlotte, NC, USA

Contents

Chapter 1
Introduction to the Digitalization of Power Systems and Markets

The first decades of the twenty-first century have seen the beginnings of a remarkable transformation in the world's energy supply and distribution systems. Renewable, distributed, and decentralized energy resources have seen exponential growth both at the grid and market penetration levels, as well as driving deregulation and liberalization. Due to the urgency of reducing carbon emissions contributing to the global climate crisis, the reign of fossil fuels as the predominant supplier of energy is rapidly coming to an end as the energy sector evolves toward a greener vision, known as the "Green Shift." The integration of digital technology from the information and communication technology (ICT) sectors into the Green Shift has triggered an additional transformation wave by adding digitalization into the picture, resulting in the "Digital Green Shift" (DGS). The DGS is a framework that aims to tackle the challenges and interrelations associated with higher penetration of renewable energy resources but at the same time making sure that the latest digitalization technologies such as artificial intelligence, blockchain technology, and advanced ICT are utilized in an orchestrated way to provide sustainable and affordable solutions. This chapter discusses the background of the Digital Green Shift, what is driving it, and what impact can be expected from this transformation in the future.

Transition of Power Systems and Digital Green Shift
Energy policy has the task of ensuring the smooth functioning of the energy sector. The goals of energy and environmental policy have changed over time as the social challenges and paradigms have changed. The United Nations has declared a set of high-level targets to deal with emerging issues related to protecting the planet and improving the well-being of humanity under the 2030 Agenda and Sustainable Development Goals (SDG) [1]. The topics discussed in this book relate to the following key goals of the UN SDG roadmap: ending poverty (goal # 1) [2], affordable and clean energy (goal # 7) [3], reduced inequalities (goal # 10) [4], sustainable cities and communities (goal # 11) [5], and responsible consumption and

U. Cali et al., *Digitalization of Power Markets and Systems Using Energy Informatics*, https://doi.org/10.1007/978-3-030-83301-5_1

production patterns (goal # 12) [6]. Achieving these goals will require massive innovation in the energy systems that support global civilization. If left to themselves, markets would probably not achieve any meaningful and sustainable solutions in a sufficient amount of time to head off global climate change and to address the other concerns in the UN SDG roadmap. As a result, achieving these goals will require government, industry, and academia to develop new technical and organizational solutions for environmentally sustainable energy systems and markets.

1.1 Multilayer Perspective Model and Energy Systems

In connection with societal transition processes, a multilevel approach that separates landscape, regime, and niche territories is helpful to understand the impact of energy systems transformation on society. Certain social subareas and functions are organized in regimes, such as energy supply. These are embedded in overarching social developments at the landscape level (e.g., climate change or new political ideologies). Innovations in the regimes (e.g., an energy supply with renewable energies) arise through developments in niches in which innovations are established that can replace existing regime structures (Fig. 1.1).

Technology transitions are the processes and events in which technological innovations, paradigm shifts, and social, economic, political, and industrial changes are happening. Multilayer perfective (MLP) is proposed by Geels [7] as an analytical framework to systematically describe how a technology transition occurs. MLP is also used to interpret and understand the interrelations between the actors, ecosystems, institutions, global events, and innovation and thereby establish a link between the technical and social sciences. MLP provides a modest heuristic methodology with which these different elements can be thought of as social subsystems.

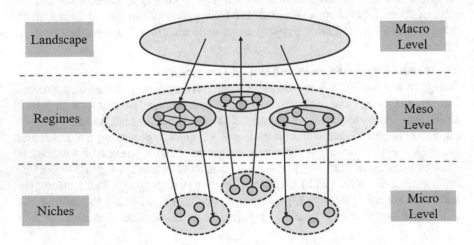

Fig. 1.1 Transition of the energy system using multilayer perfective (MLP). (Modified from [7])

However, MLP does not propose using sophisticated analysis to investigate the rise of new social movements, new scientific paradigms, or new technologies. It simply provides a framework in which comprehensive studies of these factors can be linked to other developments, which together make up the system transformation. The framework consists of three layers:

- Landscape/macro level
- Regime/meso level
- Niches: micro level

Landscape/Macro Level   Landscape is the highest level of the MLP model. This is the macro-level structural context in which each regime is rooted and where developments in the landscape can apply pressure to adjust the existing regime. However, the landscape can also stabilize the existing regime. The landscape level also acknowledges the existence and occurrence of exogenous factors that are almost impossible to influence, in addition to global rules and institutional activities that deal with global impacts. The set of rules and recommendations communicated by the World Health Organization (WHO) about combating the COVID 19 pandemic are an example. As an example, the following global trends, factors, and phenomena are relevant topics of the landscape level:

- The OPEC oil crises of the 1970s
- The Harrisburg, Chernobyl, and Fukushima reactor meltdowns
- Wars
- Natural catastrophes
- Massive demographic changes and immigration waves
- Virus pandemics
- Climate change
- New social movements
- Changes in general political ideology
- Fundamental economic changes
- New scientific paradigms
- Sociocultural developments (e.g., increasing environmental awareness)

Even though this level is named "landscape" the approach is not associated with any real spatial (country or regional) levels or political institutions and reveals nothing about how the various levels are related to actual institutional structures.

Regimes/Meso Level   This is the second level of the MLP model, and it deals with the intermediate impacts and drivers. The regime is the principal model and is used to resolve certain challenges and organize certain social subsystems and functions that are embedded and connected with some overarching global events and social developments occurring at the landscape level. Regimes are the principal models designed for resolving challenges or organizing functions and interactions in a social subarea (e.g., national renewable energy acts, institutional action plans regarding electricity supplies from various resources, etc.). Regimes are linked with the upper and lower layers. The overarching and exogenous global developments and events that are happening at the landscape level (e.g., climate change and the Fukushima earthquake and reactor meltdown) impact the regimes that react to generate new legislative actions (e.g., new political ideologies, fundamental shifts in the national energy policies resulting in the development of renewable energy laws, etc.). Innovations in the niches (e.g., new innovations in the field of renewable energy, digitalization, and energy storage) are linked to the regimes by replacing existing regime structures or influencing them dynamically.

Regimes can include the following elements:

- Guiding principles of the paradigms
- Influencing technologies and infrastructure
- Industrial structure and corporate culture
- Power market structure and the relationships between stakeholders like producers, consumers, prosumers, and aggregators
- Energy politics and regulatory frameworks

Niches/Micro Level   This is the lowest level in the MLP model where new innovations arise with a certain independence from the existing regimes in the meso layer. Niches diverge from the micro level in that they are not only about the actors but also about the interaction of the actors in a certain context, although this context differs from the regime conditions. Niches play a crucial role in transformation processes. Niches can arise "by chance" within the regime ecosystem, where the niche innovations offer solutions. Contrarily, niches can also be developed consciously and as a result of a planned strategy. Innovations can also originate within the regime layer, but these usually do not have the capability to lead to a regime transformation. For instance, innovative boilers for national gas-fired thermal power plants are being developed in the meso layer. Meanwhile, innovative and highly digitalized wind technologies and PV modules have emerged as niches in the micro layer impacted by national renewable energy legislation.

1.2   5Ds of Energy and Digital Green Shift

Energy policymakers and legislative authorities are responsible for determining regulatory policy for the energy sector. Therefore, for any type of initial market and product design-related task, it is essential to execute a comprehensive techno-political analysis that ensures the boundaries of the proposed framework are compliant with regulatory policy. The main driver of deregulation, decentralization, and decarbonization was initially driven by the OPEC Crisis in the 1970s. "Push policy" instruments based on regulation were successfully created and implemented to support renewable and distributed energy resources. During the last three decades, these policies have helped countries reach their decarbonization targets by also reducing carbon dioxide emissions. The most effective energy policy instruments have been the direct and indirect support schemes given to renewable energy investors, such as feed-in-tariffs and tax credits. The implementation of such "push policy" instruments occurs as regimes in the meso layer of the MLP model. Global trends such as OPEC Crisis, responses to climate change, Fukushima earthquake, and COVID 19 pandemics occur in the macro layer of the MLP modeling convention.

Due to technological and policy changes, electricity and financing systems have drastically evolved over the course of the last 40 years. These recent changes have allowed for increased integration of new energy resources and business models, which in return enabled consumers to take a more active part in electrical energy production. This evolution can be summarized and explained via the 5Ds [8]:

- Deregulation
- Decentralization
- Decarbonization
- Digitalization
- Democratization

Deregulation, also known as liberalization, of the electrical markets started during the global oil crisis of 1979 and transformed the classical utility model via the use of electricity markets. Therefore, large, centralized power systems and markets transformed toward smaller and inter-collaborating decentralized markets (Fig. 1.2).

The Green Shift focuses on providing more intensive support for sustainability in the energy sector by utilizing distributed and renewable energy sources (RES).

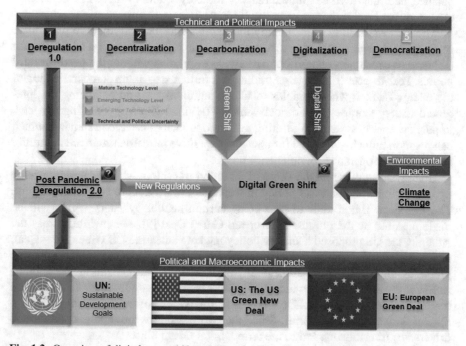

Fig. 1.2 Overview of digital green shift and 5Ds of energy

However, the variable nature of RES has led to technical challenges in the energy industry, such as imbalances between supply and demand and increased need for better forecasting techniques. Thus, smart grid systems became popular with the rise of integrated ICT and Internet of Things (IoT) capabilities, which in return allowed the digitalization of the energy industry and initiated the digital transformation of energy systems. The digital shift focuses on artificial intelligence (AI) and machine learning (ML) for providing better price, supply, and demand forecasting in addition to technical factors such as fault detection for maintenance. More recently, maturation in the fields of FinTech and tools like distributed ledger technology (DLT) can minimize transaction times for near-real-time energy trading and energy finance options, which can accelerate meeting the United Nations Sustainable Development Goals (UN-SDG).

The Digital Green Shift has an analogy with the history of the Industrial Revolution. The 1st Industrial Revolution (Industry 1.0) started with the steam engines and continued with Industry 2.0, which was mainly about substituting electricity-enabled production for steam-driven production. The 3rd Industrial Revolution was enabled and driven forward by developments in the fields of IoT and renewable and distributed energy resources. These developments led to the 4th Industrial Revolution in the energy industry due to the maturation of ICT technologies and the rise of new ones such as artificial intelligence (AI), autonomous robotics, and distributed ledger technology (DLT). Industry 4.0 is enabled by the tools of deep digitalization, which can offer advantages such as better real-time control, balance, and security. The groundbreaking Industry 4.0 innovations can be considered as micro-level niche actors in the MLP model.

The democratization of the energy market has allowed customers to participate more actively in generation as prosumers. With the further adaptation of blockchain technology, prosumers can further participate even more without the need for third parties. Peer-to-peer (P2P) energy trading is a further step in the democratization of the energy market, which enables localized solutions for grid balancing via integrated energy storage or electrical vehicles (EVs). P2P energy trading can also impact the energy system via offering mitigations such as peak load shaving, which can result in reducing the need for new and expensive infrastructure from the utility side and lower operation, maintenance costs.

Energy systems are on the verge of a second transition, which is mainly driven by international and national policy movements toward a more sustainable future. The European Union aims toward being carbon neutral by 2050, which will be made possible by the proposed European Green Deal [9]. Meanwhile, across the Atlantic, the US proposed a similar policy packet named the US Green New Deal, which includes important elements to combat climate change and socioeconomic problems. Both policies are strategically built around decarbonization and digitalization of the energy sector to incentivize increased demand for smart, efficient, and renewable energy solutions. The world is facing two major transitions in parallel: a Digital Shift and a Green Shift. Since the two transitions are acting as reciprocal drivers, the transition has led to the term Digital Green Shift.

The legislative framework for the digitalization of power systems and markets has become more complicated since it is expected to cover the overlapping and interdisciplinary domains of energy and digital technologies. For example, Germany has released a comprehensive Blockchain Strategy Document that covers energy-related use cases of blockchain in addition to other fundamental components. Other countries are expected to follow Germany's lead, as was the case with the transition to green energy (Energiewende), and to release their own national digitalization strategy in coming years that covers sector-specific components such as energy, logistics, and health legislative activities [10].

1.3 Cyber-Physical Social Systems for Energy Systems and Markets

The Smart Grid Architecture Model (SGAM) was developed as a vendor-neutral model providing a holistic view of smart grids applications. It was designed to be used as a reference architecture model under the mandate of M/490 [11]. SGAM was especially adopted in European Union countries by the industry and academia [12]. Meanwhile, the National Institute of Standards and Technology (NIST) in the US proposed a framework and conceptual reference model that has a strong emphasis on the interoperability-related issues of smart grids [13]. However, SGAM and various similar multilayer architecture models have not explicitly considered deep digitalization technologies such as artificial intelligence and blockchain technology. Therefore, it is essential to develop a multilayer reference model, which is higher resolution and considers emerging digitalization technologies as an integrated part of the next-generation smart girds framework. The model shown in Fig. 1.3 has been proposed for incorporating deep digitalization technologies into the smart grid architecture [14].

The layers in the deep digitalization smart grid architecture are the following.

Energy Policy and Regulatory Layer The energy policy and regulatory layer is responsible for developing new policies and supervising and managing energy legislation and regulation. Policymakers develop new energy legislative and regulatory policies to ensure that the energy industry satisfies the requirements of energy security and emissions for their jurisdiction. System operators, utilities, and all market participants are committed to compliance with energy policies and regulations.

Business Layer The stakeholders in the current electricity markets include power utilities, power producers, system operators of distribution and bulk power grids, trading companies, investors, among others. The diversity of power market participants has been driven by deregulation and liberalization. With the proliferation of distributed energy resources, small-scale and decentralized market participants such as prosumers and novel service providers are expected to grow quickly in the future. This new business era will be enabled by advanced ICT technologies such as DLT and artificial intelligence.

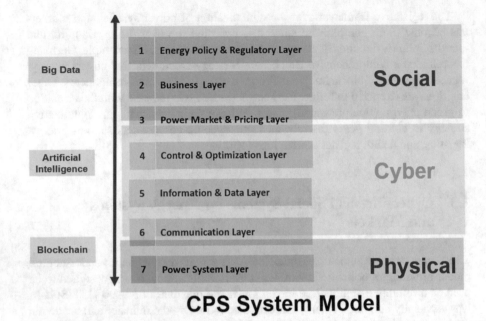

Fig. 1.3 Seven-layer cyber-physical social systems (CPSS) architecture model of power systems and markets

Energy regulatory policies, legislation, and economic metrics are considered when making investment decisions. Levelized cost of electricity (LCoE), net present value (NPV), return on investment (RoI), and discounted payback period (DPP) are among the most widely used energy economic indicators. Investors and energy economists need to perform a comprehensive cost and benefit analysis for any emerging, new technology.

Power Markets and Pricing Layer The power market is a complex platform that manages the physical power grid operation, energy trading, financial transactions and settlement, and communication and data exchange for the market participants. During the past two decades, deregulation and liberalization have transformed market rules to allow market participants, including independent power producers, non-utility power producers, and new local power markets, to trade energy and ancillary services.

Control and Optimization Layer The control and optimization layer is responsible for the physical operation of the power grid. Over the past two decades, the power industry has been adopting Supervisory Control and Data Acquisition (SCADA) systems as a standard tool to monitor system operation and make informed decisions about optimization and control of the power grid. DLT and data analytics technology such as artificial intelligence are being integrated into the existing power grid control and optimization paradigm.

Information and Data Layer Built on the communication layer, the information and data layer facilitates data processing, analysis, and cybersecurity for smart energy systems. This layer is a likely candidate for the introduction of DLT and AI infrastructure, including smart contracts and tokenization.

Communication Layer The communication layer hosts various communication protocols to facilitate interactions between different layers. Critical factors considered in the design of efficient smart grid systems include reliability, security, robustness, scalability, power consumption, and cost of the communication technologies. Chapter 2 describes the protocol standards for communication on the smart grid.

Power Systems Layer The power system layer operates the physical components of a power grid, including power generators, power transmission and distribution assets, and consumers or prosumers on the demand side.

1.4 Digitalization and Transition to Future Electrical Power Systems

The electric power demand has grown continuously to meet the electricity demand from citizens and industry since the inception of the grid in the last century. The use of renewable energy resources, that is, solar, wind, thermal, etc., and energy storage technologies, that is, batteries, has also been increasing to meet the need for decarbonized power generation. These together make the electric power grid more vulnerable, complicated, and difficult for energy suppliers to manage and balance. The integration of ICT allows the grid to meet these challenges. Digitalization uses advanced communications technology and data sources to support better decision-making, generate better analysis, and automate or control operations intelligently in each segment of the grid from generation to consumption [15].

Digitalization supported by ICT is a crucial component and enabler for driving the transformation toward more sustainable energy generation and consumption [16] by providing the necessary infrastructure for informed and intelligent management [17]. Through digitalization, the role of customers has also been changed. Customers are becoming active players in the grid and not just passive consumers, increasing their own energy awareness and, in many cases, generating power for themselves and for export to the grid at large. According to an International Energy Agency (IEA) report [18], energy systems have become more connected, intelligent, efficient, reliable, secure, and sustainable with digitalization and the use of advanced technologies in data, analytics, connectivity, and control. These digital technologies are applied to major smart grid applications, ranging from smart home/ building automation, smart metering, distribution automation, wide-area monitoring and control, and protection for a variety of smart grid use cases [19]. The most significant transformational potential for digitalization is its ability to increase flexibility and enable integration across the entire electrical power system. Digitalization can also contribute to reducing the cost of grid operations by reducing operations and maintenance costs, improving efficiency, reducing unplanned outages and downtime, and extending the operational lifetime of assets. The IEA has identified the following opportunities at the intersection of digitalization and energy: (a) smart demand response (DR), (b) the integration of renewable energy sources (RES), (c) the implementation of smart charging for electric vehicles (EVs), and (d) the emergence of small-scale distributed electricity energy resources (DERs).

The most evident effect of digitalization concerns the transport and distribution of electricity. Electronic meters constitute the automated metering infrastructure (AMI), which enables the efficient management and balance of electricity supply and demand. This solution is particularly important for intermittent renewable sources allowing them to be fully integrated with the grid but has an even greater value since it enables the grid to become increasingly flexible and decentralized. While digitalization can bring many positive benefits, it raises additional security and privacy concerns. Cyberattacks are becoming easier because IoT devices increase the "cyberattack surface" in energy systems. Energy digitalization needs to cover security, reliability, and resiliency to ensure the safety and reliability of the grid going forward.

1.5   Integrating Renewables and Storage

The major technologies behind today's renewable electricity production have been available since the mid-twentieth century but only became economical for scalable deployment in the utility industry since 2000. For example, the cost of solar photovoltaic (PV) panels [20] and lithium battery prices [21] have dropped 89% since 2010, and wind turbine costs have dropped by 50% since 2008 [22]. In addition, the power conversion efficiency available from solar PV and the storage capacity of lithium batteries has steadily increased. Solar panels are available today for residential installation in the 430 watts/panel range, while in 2004, the average panel power production was around 275 watts/panel. Lithium battery energy density has almost tripled between 2010 and 2020 [23]. Wind turbines also have become increasingly larger. Whereas 2 MW turbines were common in the early 2000s, today turbines up to 10 MW are available for offshore wind energy facilities, and 5 MW for onshore [24] Renewables coupled with storage have become so economical that some utilities are planning to replace gas-fired peak power plants, which only run for a few hours per year and are expensive to maintain, with renewables and storage.

While wind energy deployments are most suitable for large utility scale, due to the size of economically priced equipment and the need for a large amount of land for siting, solar PV and lithium battery deployments are straightforward to scale down as well as up. Like what happened with telecommunication equipment in the 1990s and early 2000s with the rise of the Internet, the decline in cost has made solar PV and lithium batteries economical for middle-class consumers to deploy, depending on the price and rate structure for electricity in their jurisdiction. This has led to the rise of the *prosumer*, a customer that both uses and generates electricity and who, in some cases, generates enough electricity in excess of their consumption to export onto the low-voltage grid for other consumers that do not have solar and storage.

Solar has become so successful in some jurisdictions that load and real-time prices for electricity have become radically depressed during sunny days when a lot of solar electricity is on the gird, while rising in the evening when the sun disappears, people return home from work, and start using appliances. This has led to the so-called "duck curve" [25], when prices on the wholesale market trend negative in the afternoon during the spring leading to curtailment of power from utility-scale solar facilities. With residential solar, a concern about backflow into the transmission network from rooftop solar has led some regulatory authorities to require prosumers to consume more of their self-generated power and release less to the distribution grid, for example, in the US state of Hawaii [26]. Residential, behind-the-meter, and utility-scale front-of-the-meter batteries can help with both these problems, storing energy when solar-generated electricity is flooding the grid and releasing it during the night and on cloudy days when solar-generated electricity is not available.

Recent research [27] indicates that accelerated deployment of solar PV and wind for generation combined with lithium batteries for storage will likely be able to meet the societal goal for complete decarbonization of the grid before mid-century in a cost-effective manner, yet there are challenges. Failure of countries to make progress on ambitious goals announced at the UN Climate Committee's meetings over the last 25 years suggests that accelerated deployment may not occur, which leads some researchers to suggest that the goal of limiting temperature increases to $1.5°$ C may be out of reach. This has led to considerations of climate engineering, primarily measures to reduce sunlight and thereby reduce heating of the atmosphere. A reduction in sunlight will cause reduced PV power generation, which will require even more panel deployment before fossil fuel power plants can be shut down.

Other renewable energy sources that have not achieved widespread deployment still seem promising and worthy of research. Geothermal electricity generation (not to be confused with geothermal space conditioning) benefits from many crossover technologies in the oil and gas industry that could contribute to making it more scalable. Yet geothermal deployments today represent a minor fraction of renewable production and are mostly limited to areas where there are proven high-temperature reservoirs relatively close to the surface, like Iceland and The Geysers in northern California. Alternative storage technologies also require study. The resources that go into manufacturing lithium batteries, while not rare, are also not overly abundant. Many research projects have demonstrated batteries based on earth-abundant materials like sodium and calcium, but the lithium battery manufacturing ecosystem is so deeply entrenched that they would have difficulty achieving traction. Long-term storage also requires additional technology development since in some cases a long period of cloudy weather or seasonally reduced sunlight can make the size of a PV system for handling load too large if it is based on the minimum generation. It would be a mistake to stop work on other promising technologies in case some political, economic, or technical barrier arises for the existing renewable technologies, or the new technologies begin to exhibit orders of magnitude cost reduction or performance improvement over existing technologies. But considering the urgency of the climate emergency, the primary focus in terms of investment and innovation needs to be on massively scaling up the deployment of existing renewable technologies and storage.

1.6   Smart and Connected VPPs and Microgrids

The trend toward replacing fossil and nuclear generation with renewables intersects the digitalization of the power industry through demand response and virtual power plant aggregators and smart microgrids. Renewables represent a generation resource that is variable both on a daily and seasonal basis and deployable in either a centralized or a highly distributed manner. Storage provides smoothing so that renewables plus storage become a dispatchable resource. Microgrids represent a possible new model for grid architecture. Digitalization is the cement binding together these disparate sources of energy and forging them into the grid of the future.

Demand response programs were originally rolled out by utilities for commercial and industrial customers in response to regulatory requirements in the US for the development of a method whereby load-serving entities could reduce the load in response to a request from the transmission system operator. Such overload conditions occur in summer when air conditioning loads exceed committed generation capacity in some markets and may start to occur in winter in colder areas as building decarbonization results in more heat pumps and other types of electric heating. A supply shortage causes the transmission system operator to institute rolling blackouts. When such a grid event is declared, the commercial and industrial customers signed up in the demand response program respond by reducing their load in various ways in return for a monetary incentive. Participation from commercial and industrial customers in demand response programs has been quite limited, and in the mid-2010s, utilities and third-party aggregators began enrolling residential consumers into the programs in an effort to broaden participation.

In parallel, the rise of DERs, especially in Europe and Australia, let to the development of virtual power plant aggregators. A virtual power plant is an aggregation of behind-the-meter renewables that can provide a firm power commitment through the use of storage. Additional measures are also possible if loads are more flexible; for example, a homeowner could precool their house during the afternoon when the grid is flooded with solar generation and then reduce or turn off their air conditioning in the evening when the grid tends to be under stress. The aggregator responds to an event from the transmission or distribution system operator by requesting action from their subscribers, either load reduction or offset from a demand response aggregator or additional power from a virtual power plant operator. An aggregator with tens of thousands of subscribers can provide sufficient flexibility to enable the grid operator to avoid rolling blackouts or, in the case of a virtual power plant operator, even contribute toward serving the distribution system's peak load directly.

Aggregators are a new development, and they have yet to reach their full potential. For example, most demand response aggregators today signal their subscribers by sending a text message to their cell phone requesting the subscriber to reduce their power consumption at a certain time the next day. Participation can be spotty even though the subscriber receives compensation, but both demand response and virtual power plant dispatching could be completely automated with the smart grid and energy IoT technologies discussed in Chaps. 4 and 5 of this book. Complete automation would improve the reliability and dispatchability of DERs, and flexible load could even lead to virtual power plant operators offering ancillary services like volt var and frequency response or replacing combined cycle gas generation plants for servicing peak load.

A microgrid is defined as a connection of generators and loads having a defined electrical boundary that allows disconnection and reconnection to the larger grid. While the generators in a microgrid can, in principle, be fossil fuel-powered, renewables such as solar and wind with lithium batteries for storage are in many cases more economical to deploy due to their lack of need for fuel. Microgrids have become more attractive as weather-related disruptions on the grid at large have increased due to global climate change. Hurricane Sandy caused massive damage to the grid in New York and New Jersey in 2012, and Hurricane Maria destroyed a large part of Puerto Rico's grid in 2017 [28, 29]. Such huge storms are a predicted consequence of the warming of the sea surface and atmosphere from global warming. In California, gigantic wildfires have resulted from the sparking of electrical lines in late summer and fall, again due to excessively dry winters caused by the reduction in rainfall because of climate change. This has led the California utilities to announce public safety shutoffs, where power is shut off to vulnerable communities when weather conditions indicate that sparking may be likely. Most such communities are small towns in rural areas near the end of transmission lines. A microgrid can help a community ride through shut off of power from the distribution system operator, or partial collapse of the main grid, as long as the microgrid distribution system remains intact. EVs can also support such kind of collapsed grid during blackouts [30].

The microgrid architecture has been incorporated into a system of systems architectural concept called the *fractal* [31] or *autonomous energy* [32] grid. Fractals are geometric objects that have the same patterns at all scales, from the smallest to the largest. Similarly, the fractal grid uses the same information and control models at all scales. Every segment has decision-making capability within its own connected domain and about whether to connect to the larger grid. Rules for when to connect and when to disconnect and how to exchange information between peer microgrids and among devices within the microgrid are defined in standards. While generation devices in the fractal grid could, in principle, be fossil-fueled, renewable devices at a small scale are a better match. The architecture is based loosely on the architecture of the Internet, which consists of autonomous systems that have local decision-making capability and well-defined rules for connecting to other networks. Such an architecture has proven to be more resilient against disruption on the Internet, as one autonomous system can fail without impacting any of the others, and this property is also advanced as a major benefit of the fractal grid.

Whether or not today's centralized grid evolves into a fractal grid is still an open question. Regulatory barriers and over 100 years of legacy design in power systems engineering are strong countervailing trends, and the technical challenges of managing grid stability in a fractal-style grid have yet to be addressed. In addition, the business arrangements regarding how today's monopoly utility business model would evolve are still unanswered. Today's Internet resulted from over 40 years of deregulation, technology, and business innovation, and reregulation in the telecommunications industry, and the electric power industry may need to undergo a similar process in order for a fractal grid architecture to take root. Nevertheless, microgrids coupled with renewables and storage can provide reliable power where the main grid is unreliable today, at scales as small as a single building. Properly coordinating the generation and load devices in a microgrid and coordinating with the larger grid all require the collection of data from the devices and communication between the main grid and microgrid. These functions are provided by the digitalization of the grid.

1.7   Summary

This chapter provides an introduction to the book, which is about the transformation of current energy systems through decarbonization and digitization supported by advanced information, communication, and control technologies besides higher penetration of RES to the power grids. The operation of the electrical power grid is in transition to a new digital era at the same time. The integration of digital technologies through the use of data analytics, connectivity, and control from ICT has changed traditional electrical power grid operations to become more intelligent, effective, and greener. The green and digital transformation of energy systems will significantly increase the overall efficiency of the energy infrastructure and reduce the cost of energy while at the same time address the global climate crisis by fully decarbonizing the world's energy supply system. Digitalization will continue to find new answers to some of the biggest challenges in the energy sector.

References

1. Transforming our world: the 2030 Agenda for Sustainable Development, https://sdgs. un.org/2030agenda, (Accessed 2021-04)
2. Goal 1: End poverty in all its forms everywhere, https://www.un.org/sustainabledevelopment/ poverty/, (Accessed 2021-03)
3. Goal 7: Ensure access to affordable, reliable, sustainable and modern energy for all, https:// sdgs.un.org/goals/goal7, (Accessed 2021-04)

4. Goal 10: End poverty in all its forms everywhere https://sdgs.un.org/goals/goal10, (Accessed 2021-04)
5. Goals 11: Make cities and human settlements inclusive, safe, resilient and sustainable, https://sdgs.un.org/goals/goal11, (Accessed 2021-04)
6. Sustainable consumption and production, https://sdgs.un.org/topics/sustainable-consumption-and-production, (Accessed 2021-04)
7. Geels, F. W., 2002. Technological transitions as evolutionary reconfiguration processes: a multi-level perspective and a case study. Research Policy 31 pp. 257-1273
8. Stekli, J., & Cali, U. (2020). Potential impacts of blockchain based equity crowdfunding on the economic feasibility of offshore wind energy investments. Journal of Renewable and Sustainable Energy, 12(5), 053307.
9. European Union, EU climate action and the European Green Deal, https://ec.europa.eu/clima/policies/eu-climate-action_en, (Accessed 2021-08-29)
10. Bunderministerium fuer Wirtschaft und Energy & Bundesministerium der Finanzen, https://www.bmwi.de/Redaktion/DE/Publikationen/Digitale-Welt/blockchain-strategie.pdf?__blob=publicationFile&v=8, (Accessed 2021-08-29)
11. CEN-CENELEC-STSI Smart Grid Coordination Group, Smart Grid Reference Architecture, November 2012, https://ec.europa.eu/energy/sites/ener/files/documents/xpert_group1_reference_architecture.pdf, (Accessed 2021-08-29)
12. M. Gottschalk, M. Uslar, C. Delfs, The Use Case and Smart Grid Architecture Model Approach: The IEC 62559-2 Use Case Template and the SGAM Applied in Various Domains (Springer, Berlin, 2017)
13. Arnold, G. , Wollman, D. , FitzPatrick, G. , Prochaska, D. , Holmberg, D. , Su, D. , Hefner, A. , Golmie, N. , Brewer, T. , Bello, M. and Boynton, P. (2010), NIST Framework and Roadmap for Smart Grid Interoperability Standards, Release 1.0, Special Publication (NIST SP), National Institute of Standards and Technology, Gaithersburg, MD, [online], https://doi.org/10.6028/NIST.sp.1108 (Accessed 2021-04-09)
14. Cali, U., & Lima, C. (2020). Energy informatics using the distributed ledger technology and advanced data analytics. In Cases on Green Energy and Sustainable Development (pp. 438–481), IGI Global.
15. The digitalization of energy brings opportunities, https://www.abb-conversations.com/2019/05/he-digitalization-of-energy-brings-opportunities/, (Accessed 2021-04-09)
16. DIGITALIZATION AND THE FUTURE OF ENERGY, https://smartenergycc.org/wp-content/uploads/2019/07/Digitalization_report_pages.pdf (Accessed 2021-04)
17. Group of Experts on Energy Efficienc, Digitalization: enabling the new phase of energy efficiency, https://unece.org/sites/default/files/2020-12/GEEE-7.2020.INF_.3.pdf (Accessed 2021-04)
18. Digitalisation and Energy, https://www.iea.org/reports/digitalisation-and-energy, (Accessed 2021-04)
19. Kuzlu, M., Pipattanasomporn, M. Rahman, S., 2014. Communication network requirements for major smart grid applications in HAN, NAN and WAN. Computer Networks, 67, 74–88.
20. TheSolarNerd.com, "Will solar panels get cheaper? (updated for 2021) – The Solar Nerd". [Online]: https://www.thesolarnerd.com/blog/will-solar-get-cheaper/ (Accessed 2021-04-12).
21. Mike Scott, Forbes, "Ever-Cheaper Batteries Bring Cost Of Electric Cars Closer To Gas Guzzlers". [Online]: https://www.forbes.com/sites/mikescott/2020/12/18/ever-cheaper-batteries-bring-cost-of-electric-cars-closer-to-gas-guzzlers/?sh=1d2568573c17 (Accessed 2021-04-12).
22. US Department of Energy, "2018 Wind Technologies Market Report". [Online]: https://www.energy.gov/sites/prod/files/2019/08/f65/2018%20Wind%20Technologies%20Market%20Report%20FINAL.pdf (Accessed 2021-04-12).
23. Kyle Field, Cleantechnica.com, "BloombergNEF: Lithium-Ion Battery Cell Densities Have Almost Tripled Since 2010". [Online]: https://cleantechnica.com/2020/02/19/bloombergnef-lithium-ion-battery-cell-densities-have-almost-tripled-since-2010/ (Accessed 2021-04-12).

24. Marcelo G. Molina and Pedro E. Mercado, ResearchGate, "Modelling and Control Design of Pitch-Controlled Variable Speed Wind Turbines", Figure 1. [Online]: https://www.research-gate.net/publication/221911675_Modelling_and_Control_Design_of_Pitch-Controlled_Variable_Speed_Wind_Turbines/download (Accessed 2021-04-12).
25. US Department of Energy, "Confronting the Duck Curve: How to Address Over-Generation of Solar Energy". [Online]: https://www.energy.gov/eere/articles/confronting-duck-curve-how-address-over-generation-solar-energy (Accessed 2021-04-12).
26. Emma F. Merchant, "Hawaii's Trailblazing Solar Market Continues to Struggle Without Net Metering", [Online]: https://www.greentechmedia.com/articles/read/hawaiis-solar-market-continues-to-struggle-without-net-metering (Accessed 2021-04-12).
27. Patrick R. Brown and Audun Botterud, "Decarbonizing the US Electricity System". [Online]: https://www.sciencedirect.com/science/article/abs/pii/S2542435120305572 (Accessed 2021-04-12).
28. Wikipedia, "Hurricane Sandy". [Online]: https://en.wikipedia.org/wiki/Hurricane_Sandy (Accessed 2021-04-12).
29. Wikipedia, "Hurricane Maria". [Online]: https://en.wikipedia.org/wiki/Hurricane_Maria (Accessed 2021-04-12).
30. Ustun, Taha Selim, Umit Cali, and Mithat C. Kisacikoglu. "Energizing microgrids with electric vehicles during emergencies—Natural disasters, sabotage and warfare." 2015 IEEE International Telecommunications Energy Conference (INTELEC). IEEE, 2015.
31. Wikipedia, "Fractal Grid". [Online]: https://en.wikipedia.org/wiki/Fractalgrid (Accessed 2021-04-13).
32. National Renewable Energy Lab, "Autonomous Energy Grids Workshop". [Online]: https://www.nrel.gov/grid/assets/pdfs/aeg-kroposki.pdf (Accessed 2021-04-13)

Chapter 2
Smart Grid Applications and Communication Technologies

The existing electric power grid architecture was built based on the energy supply and demand needs from over 100 years ago, specifically electricity flowing from large, centralized power plants to many dispersed and decentralized loads. Today with the emergence of smart grid applications, advanced information, communication, computing, and optimization technologies supporting two-way information flow, smart sensing, machine learning, and data analytics have been incorporated into the existing power grid. These technologies and applications improve grid reliability and efficiency as well as contribute to raising customer expectations. Despite such widespread deployments, the understanding of communication technology solutions that can support different smart grid applications is still unclear due to a mismatch between application requirements and utility expectations.

A smart grid is an extremely complex system, consisting of electrical and communication infrastructure with thousands of grid devices, sensors, meters, switches, controllers, distributed generators and storage devices, and many more. Communication and networking technologies play a critical role in enabling smart grid applications and manage grid devices through two-way information flow. Difficulties in developing communications and networking architecture for different smart grid applications come from the complexity and variety of different requirements. For example, the smart grid communication architecture involves heterogeneous communication technologies, generating a need for a complete analysis of the communication requirements and appropriate protocol architecture [1, 2]. Although many smart projects are implemented using existing advanced communication and networking technologies, communication technology solutions are often unclear due to gaps in smart grid standards and protocols and different communication requirements of those applications.

This chapter focuses on discussing communication network requirements for supporting different smart grid applications with respect to their reliability, latency, and payload characteristics. A comprehensive assessment is performed to evaluate different smart grid applications in generation, transmission, distribution, and

U. Cali et al., *Digitalization of Power Markets and Systems Using Energy Informatics*, https://doi.org/10.1007/978-3-030-83301-5_2

customer domains. This chapter discusses potential physical and link-layer communication technologies, including both wired and wireless, and assesses their suitability for smart grid application deployment. This chapter first describes the smart grid system and communication network architecture in Section 2.1. A comprehensive review of smart grid communication technologies is provided in Sections 2.2 and 2.3 describes different smart grid applications and their network requirements. Finally, Section 2.4 presents the chapter summary.

2.1 Smart Grid System and Communication Network Architecture

The smart grid is a dynamic platform that can be regarded as a multilayer architecture, as illustrated in Fig. 2.1. The application layer enables grid-based and customer-based applications. The security layer provides the CIA triad (i.e., confidentiality, integrity, and availability). The communication layer provides a two-way communication supporting its upper-layer applications. The power system layer supports the physical flow of electrical power from generations, transmission, and distribution to customers.

The communication layer serves as the key enabler of various smart grid applications. Different communication networks in a smart grid environment can be classified, as shown in Fig. 2.2, by their coverage range and data rate. Customer premises area networks can be classified into home area network (HAN), building area network (BAN), and industrial area network (IAN). Customer premises area networks have the lowest communication coverage (1–100 m) and data rate (1–100 kbps) requirements. Neighborhood area network (NAN) and field area network (FAN) have higher coverage (100 m to 10 km) and data rate (100 kbps to 10Mbps) requirements. Communication technologies for a wide area network (WAN) should be able to support data coverage of 10–100 km with a data rate of higher than 1 Gbps.

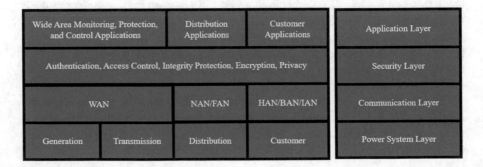

Fig. 2.1 The smart grid multilayer architecture

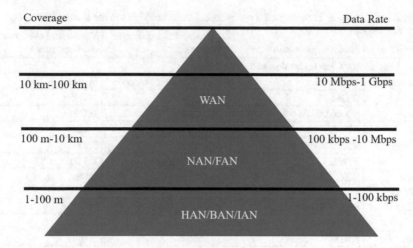

Fig. 2.2 Communication range and date rate of different networks in the smart grid communications hierarchy

2.2 Smart Grid Communication Technologies

The section explains the smart grid communication and networking technologies, both wired and wireless. The technologies with their capabilities, that is, maximum data rate and coverage range, are listed in Table 2.1. The technology description and their advantages and disadvantages are also discussed in detail. They are also briefly discussed in [3].

2.2.1 Wired Communication

2.2.1.1 Fiber-Optic Communication

Fiber optic is the most popular technology for WANs as it can support high data speed with long-distance transmission and immune to noise. It sends information over a fiber cable by turning electronic signals into light. There are several types of fiber communication that offer high data rates, that is, 40 Gbps and long-distance transmission, up to 100 km. A particular fiber communication technology is selected depending on application requirements, that is, response time, data rate, and quality of service (QoS). Different types of fiber-optic communication are explained briefly as follows:

- Passive optical network (PON) is a technology utilizing splitters to deliver data from a single optical fiber to multiple customers. A point-to-multipoint network architecture is the most viable for PON due to the efficiencies of fiber sharing and low-power consumption [4].

Table 2.1 Comparison of widely used communication technologies in the smart environment

Technology	Standard/protocol	Theoretical maximum data rate	Theoretical maximum coverage range
Wired communication technologies			
Fiber-optic	PON	155Mbps −2.5Gbps	60 km
	WDM	40 Gbps	100 km
	SONET/SDH	10 Gbps	100 km
DSL	ADSL	1–8 mbps	5 km
	HDSL	2 mbps	3.6 km
	VDSL	15–100 mbps	1.5 km
Cable Internet	DOCSIS	172 mbps	28 km
PLC	Broadband	14–200 mbps	200 m
	Narrowband	10–500 kbps	3 km
Ethernet	802.3x	10Mbps-10Gbps	100 m
Wireless communication technologies			
Z-Wave	Z-Wave	40 kbps	30 m
ZigBee	ZigBee	250 kbps	100 m
	ZigBee Pro	250 kbps	1600 m
Wi-Fi	IEEE 802.11a	2 mbps	
	IEEE 802.11b	11 mbps	
	IEEE 802.11 g	54 mbps	
	IEEE 802.11n	600 mbps	100 m
	IEEE 802.11 ac	6933 mbps	
	IEEE 802.11ax	9607 mbps	
WiMAX	802.16	75 mbps	50 km
Wireless Mesh	Various (e.g., cellular, 802.11, 802.15, 802.16)	Depending on protocols	Depending on deployment
Cellular	2G	14.4 kbps	
	2.5G	144 kbps	
	3G	2 mbps	50 km
	3.5G	14 mbps	
	4G	100 mbps	
	4.5G	1 Gbps	7 km
	5G	250 mbps/3 Gbps	2 km/50 km
LoRa	LoRa	50 kbps	5 km/20 km
NB-IoT	LTE Cat NB1	200 kbps	100 km
LTE-Cat	LTE-Cat-0	1 mbps	
	LTE-Cat-1	10 mbps (downlink)/5 mbps (uplink)	18 km
	LTE-Cat-M	4 mbps	
Satellite	Geosynchronous Satellite Internet	25 mbps downlink/3 mbps uplink	Worldwide
	Low Earth Satellite Internet	1 Gps	Planned worldwide
	GPS	50 bps	Worldwide

- Wavelength division multiplexing (WDM) is the most widely deployed technology for high-capacity, long-distance fiber-optic communication because it supports a high bandwidth capacity. Several data streams are transmitted over the same fiber-optic line through the use of multiple wavelengths, thereby reducing the number of fibers in WDM networks [5].
- Synchronous optical networking (SONET)/synchronous digital hierarchy (SDH) is a standardized time-division multiplexing (TDM) technology to carry high-volume data over a fiber-optic cable. SONET technology is deployed in the USA, Canada, and Japan, while SDH is deployed elsewhere. SONET and SDH only differ in their asynchronous bit rates [6].

Fiber-optic communication technologies are widely used to provide backhaul networks supporting a variety of smart grid applications that require reliable communication and high data transmission rate. However, their upfront investment and maintenance costs are higher than other communication technologies.

2.2.1.2 Digital Subscriber Line (DSL)

DSL is a wired communication technology that transmits data using existing telephone lines [7]. There are three forms of DSL:

- Asymmetric DSL (ADSL) allows more bandwidth for downstream than upstream data, that is, downstream at up to 20 Mbps and upstream at up to 1.5 Mbps [8]. ADSL is more suitable for use cases where more data is download than upload.
- High-speed DSL (HDSL) provides up to 2.048 Mbps data rate.
- Very high data rate DSL (VDSL) provides yet higher data transmission than others, that is, up to 100 Mbps.

DSL uses copper wire for the physical communication medium, providing data transmission up to a distance of 5 km. The main advantage of DSL is that it can be deployed quickly as already a large number of customers have access to telephone services. This makes DSL a potential communication option for many smart grid applications that require access to homes, such as smart metering. The major disadvantage of DSL is the degradation of data signal quality according to the distance, resulting in a lower data rate for longer distances. Therefore, DSL is not suitable for customers who live a long distance away from a service provider nor for critical smart grid applications.

2.2.1.3 Cable Internet

Cable Internet is a data transmission technology utilizing the existing television coaxial cable networks to provide data communication for end-users. Data over cable service interface specification (DOCSIS) provides a high transmission rate, that is, up to 172 Mbps, with a coverage range of up to 28 km. It uses the existing cable infrastructure, that is, hybrid fiber-coaxial (HFC) [9]. DOCSIS cable Internet services can provide the communication link between customer premises and a distribution system operator for metering, demand response, and home energy management applications. The main drawback of cable Internet is that the entire bandwidth on a cable segment is shared among different users. Hence, the maximum attainable data rate for any customer depends on the number of subscribers in the area.

2.2.1.4 Power Line Communication (PLC)

Power line communication (PLC) uses readily available electrical transmission/distribution infrastructures to transmit data [10]. In PLC, the data to be transmitted is modulated into a high-frequency carrier, which is then injected into the power line. PCL is cost-effective as it uses the existing power line infrastructure for communication. PLC is an ideal communication option for many smart grid applications, including smart metering, demand response, energy management, home, and building automation, plug load and lighting control, HVAC control, streetlight control, and many more. Additionally, it is the best communication candidate for a rural area that does not have other communication infrastructures. However, there are significant technical concerns with PLC due to noisy environments and harsh conditions in the power line environment. These make it difficult to ensure data quality for PLC.

2.2.2 Wireless Communication

2.2.2.1 ZigBee

ZigBee is a low-power, cost-effective, and short-range wireless communication protocol (IEEE 802.15.4). It runs on a variety of unlicensed Industrial, Scientific, and Medical (ISM) frequency bands, that is, 868 MHz, 915 MHz, and 2.4 GHz. ZigBee data rates vary according to its operating frequency band. The data rate can be up to 250 kbps at 2.4GHz, up to 40 kbps at 915 MHz, and up to 20 kbps at 868 MHz. ZigBee was originally designed for short-range wireless communication, up to 100 meters. However, the coverage range of some newer ZigBee versions, such as ZigBee Pro, can be up to 1600 meters. ZigBee has its use in premises network applications, plug load and lighting control, HVAC control, energy monitoring, industrial plant management, smart metering, and many more. ZigBee supports various network topologies, such as star, tree, and mesh, with the mesh being the most popular in ZigBee networks. Using 128-bit AES encryption, ZigBee networks are secured and can provide a robust security layer [11]. ZigBee is suitable for smart home/ building applications because it propagates, especially in the lower frequency bands resulting in good connectivity over its range. ZigBee technology also handles severe interference problems with other networks using the same spectrum, such as Wi-Fi.

2.2.2.2 Wi-Fi

Based on the IEEE 802.11 series of standards, Wi-Fi supports a variety of data rates, that is, 802.11b-11 Mbps, 802.11 g-54 Mbps, 802.11n-600 Mbps, 802.11 ac-6933 Mbps, and 802.11ax-9607 Mbps [12]. Wi-Fi operates in several unlicensed ISM frequency bands, specifically 2.4 GHz, 3.5 GHz, 5 GHz, and 6 GHz [13]. Wi-Fi can support data rates from 2 Mbps to 1 Gbps, and a typical Wi-Fi data transmission distance is up to 100 meters. It is another promising wireless technology for smart grid applications, providing reliable, secure, and high-speed data communication. However, Wi-Fi is designed for short-range wireless communications up to 100 meters. This reduces its potential for smart grid applications that require a longer range. Additionally, Wi-Fi products cost more and consume more power than ZigBee and Z-Wave.

2.2.2.3 Wireless Mesh

Wireless mesh is not a connectivity technology but rather a network architecture that can be used with a variety of wireless connectivity technologies to build a flexible wireless network. Wireless mesh nodes consist of mesh clients, routers, and gateways. Each of these nodes can serve as a repeater. They are working together to choose the quickest route for data transmission, known as dynamic routing. One of the key advantages is that communications in a large area can be served by connecting nodes in a wireless mesh fashion. In addition, it is cost-effective and has a self-healing feature [14]. Mesh networks can be deployed using cellular, 802.11, 802.15, 802.16, and more. These characteristics make wireless mesh networks a good candidate for many smart grid applications. However, wireless mesh networks suffer from interference with wireless communication technologies.

2.2.2.4 Z-Wave

Z-Wave is a proprietary short-range wireless communication technology that is low-power and low-cost. Z-Wave has its use in smart home/building applications [15]. Using the 900 MHz ISM frequency band, Z-Wave also supports mesh networks. However, its range is limited (up to 30 m). It has a low data transmission rate, that is, up to 40 kbps, but with much lower power consumption compared to Wi-Fi and ZigBee [16].

2.2.2.5 WiMAX

WiMAX (Worldwide Interoperability for Microwave Access) is one type of 4G technology. Based on the IEEE 802.16 series of standards, WiMAX is designed for metropolitan area networks (MAN). It operates a variety of frequency bands, that is,

2.3/2.5/3.3/3.5 GHz (licensed) and 5.8 GHz (unlicensed). It provides high-speed broadband service at 75 Mbps and can cover the area within a radius of 50 km. WiMAX is a perfect candidate for smart grid applications, such as monitoring and control of transmission and distribution networks, smart metering, etc. The main advantage of WiMAX is that it provides high-speed and long-distance wireless data communication that can support thousands of customers with 8000 square km in a coverage area of a base station. Nevertheless, WiMAX is an expensive communication technology requiring high power consumption [17].

2.2.2.6 Cellular

Wireless cellular technology has been deployed widely by network operators to provide consumers with cellular telephony and mobile Internet service. Cellular technology economizes on the spectrum by reusing frequencies within different cells of varying sizes depending on the frequency band within a wide geographical area and can be broken into five generations with intermediate versions, that is, 1G, 2G (GSM), 3G (UMTS), 4G (WiMAX and LTE), and 5G with intermediate versions, that is, 2.5G (GPRS and EDGE), 3.5G (HSPA), and 4.5G. 2G and 3G technologies have been started to be phased out, and 4G and 5G supporting higher data rates to become widely available worldwide. Cellular systems commonly operate in a variety of frequency bands ranging from <1 GHz/3/4/5 Ghz/24–28/37–40/64–71 GHz to 95 GHz. Existing cellular communication services can be a good candidate for supplying network access to smart grid applications, such as smart metering and AMI. Newer generation cellular technologies like 3/4/5G run at higher speed and have lower latency, are secure, and provide long-distance wireless communication technology. 5G offers the highest data transmission speed, typically 20 times faster than the 4G networks, and the lowest latency, less than a millisecond [18]. The accessibility of currently available commercial cellular towers provides some additional advantages, that is, rapid deployment and cost-effectiveness. Nevertheless, cellular technology has a higher start-up and maintenance cost than other wireless communication technologies since subscription rates can be expensive. In addition, simultaneous use of cellular services with other mobile subscribers may cause communication speed to drop significantly during periods of high use.

2.2.2.7 LoRa

LoRa is one of the trending low-power and long-range communication and networking technologies. LoRa operates in several unlicensed ISM bands, that is, 868 MHz in Europe, 915 MHz in North America, and 433 MHz in Asia. LoRaWAN is a low-power, wide-area LoRa-based communication protocol. LoRaWAN has been used for smart grid applications for smart metering, monitoring field devices through IoT-based networks, etc. Networked using a star topology, LoRaWAN requires a gateway to enable communication between a base station and its devices. LoRa can provide a data rate of up to 50 kbps for up to 20 km in rural and 5 km in urban areas [19].

2.2.2.8   Narrowband-IoT (NB-IoT)

Narrowband-IoT (NB-IoT) is one of the trending LPWAN technologies released by 3GPP, especially for enabling mobile IoT applications requiring low-cost, low data rates, and long battery life. It is based on the long-term evolution (LTE) protocol with reduced LTE protocol functionality. It uses the LTE licensed frequency bands [20] range and occupies a single narrow band 200 kHz wide. With NB-IoT, a dedicated gateway is not required, thus allowing end devices to connect with a base station directly. NB-IoT can provide a data rate of 26 kbs in Release 13 of the 3GPP standard and 127 kbps in Release 14. A coverage distance of NB-IoT is now up to 100 km [21]. NB-IoT technology is a good candidate for supporting smart grid applications with small data packet sizes, such as metering, smart street lighting, etc.

2.2.2.9   LTE-M

Long-term evolution category M (LTE-Cat-M, LTE-Cat-M1, or LTE-M) is a low-power, wide area network (LP-WAN) standard released by 3GPP (the group behind 3G) that supports IoT applications. Traditional LTE networks can provide high data rates over a long distance but consume too much power. This makes LTE not suitable for IoT applications requiring low complexity, such as low-cost devices with a long battery lifetime. 3GPP has released several LTE-based technologies. The first-generation LTEs are LTE Category 0 (LTE-Cat-0) and Category 1 (LTE-Cat-1). LTE Category M (LTE-Cat-M1) is considered as the second-generation technology designed to overcome the complexity and high-power consumption issues. It differs from the first-generation technology in several ways, specifically in lower data rate and lower power consumption. LTE-M technology allows end-user devices to communicate with an LTE network without a gateway [22]. It uses a licensed spectrum. The LTE-M frequency spectrum defined by 3GPP contains 29 frequency bands ranging from 400 MHz, 700 MHz, 900 MHz to 2.1 Ghz. Over 80% of those frequency bands allocated for LTE-M are in the below 2 GHz range. LTE-M can provide a data rate of 1 Mbps for 3GPP Release 13 and 4 Mbps for Release 14 with a coverage distance of 18 km [23].

2.2.2.10   Satellite Communication

Satellite communication provides an alternative method for bidirectional Internet communication via a link through a satellite in geosynchronous or low Earth orbit. An advantage of satellite Internet service is that it is typically available worldwide, in contrast with wired and wireless WAN service that is restricted to areas where service providers have deployed connectivity equipment. In addition, satellites are widely used for positioning and location.

Geosynchronous satellite Internet has been commercially available for a number of years and provides global coverage through ground stations, some of which are quite compact, which connect the end device to the satellite via radio. End devices connect to the ground station via standard wired and wireless Ethernet protocols. Geosynchronous satellites have a downlink bandwidth of 25 Mbps and an uplink bandwidth of 3 Mbps with a latency of 638 ms. The high latency comes from the long-distance and corresponding speed of light delay between the Earth's surface and geosynchronous orbit, around 35,786 km.

Recently, some companies have started deployment of satellite constellations in low Earth orbit, from 160 to 1000 km for low Earth orbit satellite Internet. As with geosynchronous satellite Internet, this service requires a separate ground station, which, as with geosynchronous satellite Internet, is quite compact, that the end devices connect through using standard link technologies. Initial beta deployments of low earth service have latencies of 25–35 ms and data rates of 60 Mbps downlink and 17.7 Mbps uplink, but are promising 1 Gbps when the constellation is fully deployed.

Satellite positioning is another service that has been commercially available for a number of years. There are six different positioning systems in operation currently, launched by the US, European Union, Russia, China, Japan, and India, of which the Global Positioning System (GPS) from the US is best known. These systems do not offer bidirectional communication but only a unidirectional signal from which a receiver can determine their location and the current time. The satellites are in medium earth orbit (above 20,000 km). The receiver uses the triangulation of signals from a minimum of four satellites to compute its location.

For smart grid applications, satellite communication is especially applicable if there is no available communication infrastructure in areas such as for equipment located in remote areas. However, bidirectional Internet connectivity, especially for geosynchronous satellites, suffers from long round-trip delays and the high cost of the base station [24]. In addition, some satellite links are in frequency bands where interference from heavy rain may disrupt the service, though light rain or fog should not pose a problem.

2.3   Smart Grid Applications and Network Requirements

This section describes major smart grid applications in the generation, transmission, distribution, and customer domains [25]. Figure 2.3 shows a mapping between smart grid application categories and the specific power system domain they serve.

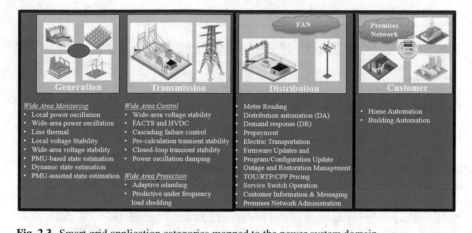

Fig. 2.3 Smart grid application categories mapped to the power system domain

2.3.1   Premises Network Applications

A premises network supports communications among various devices on the customer premises (home, office building, factory, etc.), such as sensors, controllers, and energy management units. A home area network (HAN) provides communications within a house, allowing different devices with energy control capability, such as smart thermostats, sensors, and electric vehicles, to communicate with its home energy management (HEM) unit. On the other hand, a building area network (BAN) and an industrial area network (IAN) provide communications for applications in commercial and industrial buildings, such as heating, cooling, and lighting control, and industrial energy management. These premises area networks can connect to smart grid control points, such as an electric utility/distribution system operator or an aggregator, through a communication gateway. This allows an electric utility to monitor and send signals to selected devices in HAN/BAN/IAN.

Various communication technologies can be deployed in a premises area network that requires minimal data transmission rate and coverage range. Important network requirements for applications in premises area networks are low power consumption, low initial cost, and security. The data transfer requirements for premises network applications are listed in Table 2.2.

2.3.2   Neighborhood Area Network Applications

A neighborhood area network (NAN) enables the collection of data from customers to an electric utility and the distribution of price or control signals from an electric utility to customers. NAN is called a field area network (FAN) when referring to the connection between field devices and a utility. NAN/FAN supports applications in

Table 2.2 Network system requirements for HAN/BAN/IAN applications in the smart grid

Application	Typical data size (bytes)	Typical data sampling requirement	Latency	Reliability
Home automation	10–100	Once a variety of time periods (e.g., 1 min, 15 min, hour, day, month, configurable period, etc.)	Seconds	>98%
Building automation	10–100	Once a variety of time periods (e.g., 1 min, 15 min, hour, day, month, configurable period, etc.)	Seconds	>98%

an electric power distribution system, such as distribution automation, fault detection, isolation and restoration, and Volt/VAR (Volt-Amps Reactive) control. It also supports applications that involve the interaction between electric power distribution and customer systems, such as remote meter reading, demand response, real-time pricing, and outage management. NAN/FAN is interfaced with WAN through a core network, where data are aggregated and transferred between smart grid applications and the utility data storage, billing, and monitoring services.

Smart grid applications in NAN/FAN require higher data rates and a larger coverage range than those in premises area networks. These applications are briefly discussed below, and their communication network requirements are summarized in Table 2.3.

2.3.2.1   Meter Reading

Smart meter reading is the most adopted application in a smart grid. Meter reading and related other applications facilitate collecting customer usage data from different utility services, like electricity and gas, and transporting it to a central database for data management and analysis. Utilities can collect usage data and transport the collected data to a centralized data management system where it can be analyzed, both for real-time control and for historical purposes such as billing and facilities planning. This improves meter reading and forecasting accuracy, as well as reduces operational and maintenance costs. Additionally, customers can be notified regarding their instant and total energy usage, allowing them to better manage their electricity consumption. There are three types of metering service: (1) on-demand meter reading – reading is performed if needed based on a request from the customer and the utility at any time; (2) scheduled meter interval – the meter stores usage information at a fixed time interval automatically at the customer site, which is then retrieved by the utility; and (3) bulk transfer reading – the AMI head-end system at the utility collects usage information from all meters within a NAN and transports the usage data to a meter data management system (MDM) every 24 h.

Depending on metering service types, network requirements vary greatly. Typical meter reading data message size is around 100 bytes for on-demand reading, can be up to 2400 bytes for scheduled interval reading, and can be up to several megabytes for bulk transfer. Latency can range from 15 s for the on-demand reading to 4 h for scheduled interval meter reading service.

Table 2.3 Network system requirements for smart grid applications in NAN

Smart grid application	Typical data size (bytes)	Data sampling intervals	Latency	Reliability
Meter reading				
Bulk transfer	xMB	x per day for a group of meters(from 6 am to 6 pm)	< 1 h	> 99.5%
Scheduled interval	1600–2400	4–6 per meter per day (24 × 7)	< 4 h	> 98%
On-demand	100	25 out of 1000 meters per day (from 7 am to 10 pm)	< 15 s	> 98%
Pricing				
TOU	100	1 per device per price data broadcast event – 4 per year (24 × 7)	< 1 min	> 98%
RTP	100	1 per device per price data broadcast event - 96 per day (24 × 7)	< 1 min	> 98%
CPP	100	1 per device price data broadcast event – 12 per year (24 × 7)	< 1 min	> 98%
Prepaid electricity service				
Prepayment program messaging to customer	50–150	25 per prepaid meter per month (from 7 am to 10 pm)	< 30 s	> 98%
Demand response				
DLC	100	1 per device per broadcast request event (24 × 7)	< 1 min	> 98%
Service switch operation				
Service switch operation	25	1–50 per 1000 per electric meter per day (from 8 am to 8 pm)	< 1 min	> 98%
Distribution automation (DA)				
Field DA maintenance (data acquisition)	50–1000	1 per device per 5 s - 1 per device per 12 h (24 × 7)	< 5 s	> 99.5%
Volt/VAR	150–250	Open/close CBC: 1 per CB per 12 h Open/close switch: 1 per switch per week Step up/down VR: 1 per VR per 2 h (24 × 7)	< 5 s	> 98%

(continued)

Table 2.3 (continued)

Smart grid application	Typical data size (bytes)	Data sampling intervals	Latency	Reliability
DSDR	150–1000	Sensor data acquisition: 1 per device per 15 min Open/close CBC: 1 per CB per 5 min Step up/down VR: 1 per VR per 5 min Open/close SW: 1 per switch per 12 h (1–6 h duration, 4–8 times a year)	< 4 s	> 99.5%
FLISR	50	1 per device per isolation step event (within <1 min of fault event)	< 5 s	> 99.5%
Outage and restoration management				
ORM	25	1 per meter per event (24 × 7)	< 20 s	> 99.5%
Distribution customer storage				
Dispatch	50	288 per device per day (24 × 7)	< 5 s	> 99%
Islanded	50	1 per device per detected transformer loss of power event (24 × 7)	< 4 s	> 98%
Electric transportation applications				
Energy supplier sends price info to PHEV	255	Per charging event per day (from 7 am to 10 pm)	< 15 s	> 99.5%
Energy supplier interrogates PHEV charge status	100	2–4 per PHEV per charging event per day (7 AM – 10 PM)	< 15 s	> 99.5%
Firmware updates				
Firmware updates	400 k–2000 k	2 per meter per year (24 × 7)	< 4 h per meter	> 98%
Program/configuration update	25 k–50 k	1 per meter per broadcast event (24 × 7)	< 3 days per 100,000 meters	> 98%
Customer information and messaging				
Send information to ODW	xMB	x per day (groups of meters) (from 10 am to 6 pm)	< 1 h	> 98%
Receive information from the utility	200	100 per 1000 per meter per day (from 7 am to 10 pm)	< 15 s	> 98%
Premise network administration				
Premise network admin	25	x per device per join request per year (24 × 7)	< 20 s	> 99.5%

2.3.2.2 Pricing Applications

In smart grid pricing applications, a pricing service broadcasts price information to customer meters and smart devices at customer premises, depending on the pricing program. Three types of well-known pricing programs are time-of-use (TOU), real-time pricing (RTP), and critical peak pricing (CPP). In a TOU scheme, the energy price changes based on a particular time period. TOU pricing programs are designed to encourage customers to reduce energy demand during peak periods and shift load to off-peak hours [26]. In an RTP scheme, the energy price is decided by the pricing service overtime periods, for example, every 5 min, 30 min, or hour, allowing local controllers on the customer premises to reduce energy during high price periods. In a CPP scheme, energy prices are increased during times of high peak demand, like in the evening on the hottest summer days, to drastically curtail loads. CPP is the most restricted program, and the electricity price increases substantially during a CPP event to encourage load reduction. A typical size for RTP/CPP/CPP message is 100 bytes with the required latency of less than 1 min.

2.3.2.3 Prepaid Electricity Service

Prepaid electricity service allows customers to pay for their electricity usage in advance [27] rather than paying for the usage at the end of the month. Customers get service as long as they have some credits in their account. A typical prepaid electricity service measures electrical usage and deducts the cost from the credit in the account in real time. The customer is notified through email or text message, or the meter issues a warning like an audible alarm or warning messages once the credit reaches a threshold. Customers need to refill the account to gain service again. Otherwise, their electricity service is disconnected after a certain amount of time. A typical data message size for the prepaid electricity service is between 50 and 150 bytes with a latency of less than 30 s.

2.3.2.4 Demand Response (DR)

DR is a voluntary program providing an opportunity for customers to participate in distribution grid load reduction. DR allows customers to cut down or defer their energy usage during peak to off-peak periods in exchange of a financial incentive, thus lowering electric bills. DR programs are either direct load control or price-based program, such as ToU, RTP, and CPP [28]. Direct load control programs (DLCs) are the simplest and most popular of the load-responsive DR programs. Typical appliances involved in DLC programs are central HVAC systems, electric water heaters, and pool pumps. For a typical DLC program, the selected loads are disconnected by a load controller in response to a DR signal sent by a utility during a DR event. A DR signal can be a unicast, multicast, or broadcast message depending on the DR program and the amount of load reduction required. For DR applications in general, the typical data message size is 100 bytes with a latency of less than 1 min.

2.3.2.5 Service Switch Operation

Service switch operation is an example of a smart grid application in NAN/FAN. Service switching allows an electric utility to switch its service on or off without sending a service truck to the customer site. With the service switch application, a utility can reduce time and effort to terminate a utility service. Service switching is performed by sending a service switch enabling/disabling command to a meter and requesting the operational status of the meter service switch. For service switch operation, the typical data message size is 100 bytes with a latency of less than 1 min.

2.3.2.6 Distribution Automation (DA)

DA is the most important smart grid application for improving service reliability, operating efficiency, and minimizing. DA provides real-time monitoring and management of electric distribution assets such as capacitor banks, voltage regulators, line sensors for voltage and current, and switches [29]. DA applications and their requirements change depending on the DA use cases. Four common DA use cases are

- Distribution grid monitoring and maintenance, which involves reading data from field devices, such as sensors, meters, relays, and other intelligent electronic devices (IED), to monitor DA assets. The collected data is used to increase grid efficiency and minimize system downtime.
- Vol/VAR control aims to improve the efficiency of energy supply and reduce energy loss through voltage adjustment and power factor regulation on a distribution line.
- Conservation voltage reduction (CVR) aims to reduce distribution-level voltage to the minimum required level to help reduce end-use energy consumption.
- Fault location, isolation, and restoration (FLISR) detects fault locations, isolates the fault on a distribution grid, and restores power after a fault occurrence to minimize customer minimal outage time.

The DA system monitors a variety of distribution equipment, including electrical lines and other devices, as well as an IP-based communications backbone with high speed and low latency. Typically, distribution management systems (DMS) and SCADA systems are integrated with the DA system. A typical data message size for DA applications is 50–1000 bytes with a latency of less than 5 s. In addition, some DA applications such as DSDR and FLSIR require high reliability of more than 99.5% due to their impact on the grid reliability.

2.3.2.7   Outage and Restoration Management

Utilities use outage and restoration management (ORM) to manage the grid during an outage event. An outage is detected when power is lost at customer premises devices, such as smart meters. ORM applications also allow an outage management system (OMS) to confirm the success of a restoration activity by providing power status verification. OMS sends power status requests to an individual meter or a number of meters, service points, or distribution nodes. Responses from devices include power status messages indicating whether the power is on or off and communication failure information. For ORM applications, the typical data message size is 25 bytes with a latency of less than 25 s.

2.3.2.8   Distributed Storage

Distributed storage applications have become more popular as technological solutions with the increasing renewable penetration. These applications involve the use of energy storage devices, such as batteries located on distribution feeders or laterals, to improve efficiency and reliability of the electricity supply through the peak load shaving, demand control, and interruption protection. Note that this application is distinct from behind the meter storage, where the storage is installed on the customer premises and is owned by the customer, in that for distributed storage, the batteries are owned and managed by the utility for grid stability purposes. Distributed storage can be used in either dispatch or islanded applications. Dispatch applications include discharging storage devices to perform peak shaving, voltage support, and demand control. Islanded applications provide temporary power to customers during an outage in a distribution feeder circuit. For distributed storage applications, the typical data message size is 50 bytes with a latency of less than 5 s.

2.3.2.9   Electric Transportation

Electric transportation is one of the most attractive smart grid applications and includes vehicle-to-grid (V2G) and grid-to-vehicle (G2V) exchange modes. V2G applications provide electricity from vehicles to the grid, while G2V provides electricity flow in the opposite direction, from the grid to vehicles. These applications also are compatible with a variety of electric vehicle drive train technologies, such as hybrid, plug-in hybrid, and full-electric vehicles. In addition, they enable EVs as a mobile distributed generation. In a typical electric transportation application, a utility sends control commands/price signals to a collection of EVs based on current power needs and also sends inquiries to EVs to obtain their battery state of charge for electric transportation applications. The typical data message size is 100 bytes, and the latency is less than 15 s.

2.3.2.10   Firmware Updates and Change Program/Configuration

Firmware updates and change program/configuration applications are used by utilities to adapt their device to new functionalities, features, and settings. The former is performed to fix bugs or add features in the firmware, while the latter updates settings, operating systems, and application software. Firmware updates and change program/configuration are usually performed due to changes in application requirements and requirements to reinforce device-level and system-level security. For firmware update applications, the typical data size is 400–2000k bytes with a latency of up to 4 h, while those for program and configuration update applications are 25–50 Kbytes with a latency of up to 3 days.

2.3.2.11   Customer Information and Messaging

Customer information and messaging applications are the most interactive with customers, allowing them to access account information, past energy usage, enrolled DR and pricing program, outage information, etc. That information is typically stored on a secure database operating by the utility, and customers can access their information stored on the database through a web portal. For customer information and messaging applications, the typical data message size is 200 bytes with a latency of less than 15 s.

2.3.2.12   Premises Network Administration

Premises network administration applications are used to connect/disconnect customers' devices to/from a network through a web portal. For premises network administration, the typical data message size is 25 bytes, and the latency is less than 20 s.

Network requirements for smart grid NAN/FAN applications may vary depending on the application, as discussed in the examples above. Table 2.3 summarizes network requirements for the above NAN/FAN applications. According to Table 2.3, these applications require higher data rates, lower latency requirements, and more security than HAN/BAN/IAN applications. In addition, satellite communication may also serve for NAN/FAN application to support remote monitoring in areas where other wired and wireless communications options are not available.

2.3.3   Wide Area Network (WAN) Applications

WAN data transmission links are typically deployed to support backbone data transmission. WAN supports communications between a utility control center and power generation plants, transmission networks, substations, protection circuits, phasor measurement units (PMUs), and data aggregation units in NAN/FAN. WAN communication and networking technologies supporting a significantly higher data rate of up to Gbps and can provide large area coverage of up to 100 km. Physical layers based on fiber optic, WiMAX, and cellular technologies are appropriate in these circumstances. Satellite communications can also serve as a backup communication link for critical network components.

WAN applications aim to handle real-time monitoring, control, and protection of the power grid to prevent cascading failures. WAN applications are generally classified into three domains: wide area control, wide area monitoring, and wide area protection (WAMPC). WAMPC system can be identified as a state-of-the-art solution of power systems, which is centralized data processing system to improve system reliability, operation, and protection, as well as to ensure system stability with the acceptable system frequency and grid voltages. As compared to traditional supervisory control and data acquisition (SCADA) systems, a WAMPC system has a significantly higher data rate and quicker latency requirements. In particular, WAMPC applications require data resolution of milliseconds. In contrast, the data resolution of SCADA is every 1 or 2 min. The subsections below discuss the three application classes of WAMPC, while Table 2.4 summarizes the requirements.

2.3.3.1   Wide Area Control

Wide area control systems (WACSs) provide real-time control of electrical devices at the electric transmission level to enable automatic self-healing capabilities and provide transient stability support. WACS offers control of fast-acting equipment like flexible-AC transmission system (FACTS) devices. It also provides other grid stabilization services, such as control of voltage stability, transient stability, power oscillation damping, and frequency stability.

2.3.3.2   Wide Area Monitoring

Wide area monitoring systems (WAMSs) collect data in real time from phasor measurement units (PMUs) and others. PMUs are single microcontroller-based measurement units that capture time-synchronized voltage and current with the phase angle of a power grid. WAMS applications are power frequency oscillation, line thermal conditions, and system voltage stability monitoring, as well as PMU-based state and dynamic state estimation. WAMSs improve operators' situational awareness and system operations in terms of reliability, stability, efficiency, and security by providing information and detecting faults, in most cases in real time, about power system stability issues.

Table 2.4 Network requirements for wide area monitoring and wide area protection (WAMPC) applications

Application	Typical data size (bytes)	Data sampling intervals	Latency	Reliability
Wide area control				
Voltage stability control	18	0.5–5 s	< 5 s	> 99.9%
FACTS and HVDC control	18	0.5–2 min	< 2 min	> 99.9%
Cascading failure control	18	0.5–5 s	< 5 s	> 99.9%
Transient stability control	18	0.5–2 min	< 2 min	> 99.9%
Closed-loop transient stability control	18	0.02–0.1 s	< 0.1 s	> 99.9%
Power oscillation damping control	18	0.1 s	< 0.1 s	> 99.9%
Wide area monitoring				
Local power oscillation monitoring	52	0.1 s	< 30 s	> 99.9%
Power oscillation monitoring	52	0.1 s	< 0.1 s	> 99.9%
Line thermal monitoring	52	1–30 s	< 5 min	> 99.9%
Local voltage stability monitoring	52	0.5–5 s	< 30 s	> 99.9%
Voltage stability monitoring	52	0.5–5 s	< 5 s	> 99.9%
PMU-based state estimation	52	0.1 s	< 0.1 s	> 99.9%
Dynamic state estimation	52	0.02–0.1 s	< 0.1 s	> 99.9%
PMU-assisted state estimation	52	0.5–2 min	< 2 min	> 99.9%
Wide area protection				
Adaptive islanding	18	0.1 s	< 0.1 s	> 99.9%
Predictive under frequency load shedding	18	0.1 s	< 0.1 s	> 99.9%

2.3.3.3   Wide Area Protection

Wide area protection systems (WAPSs) are one of the most critical WAM applications since they provide fully automated protection of power systems against unexpected cascading blackouts. WAN applications generally involve large-scale load shedding, reactive power balancing, and islanding to achieve stability if an unbalanced condition is predicted. WAPS, on the other hand, deals with unexpected contingencies that require a rapid response time to prevent blackouts. Hence, WAPS requires a short response time, that is, milliseconds up to minutes, and very high communication reliability.

In a WAMPC system, data from time-synchronized devices like PMUs, including measurement values, timestamps, and device status, are collected from different units and locations, with a high sampling rate over a WAN. The IEEE Standard for Synchrophasors for Power Systems, that is, IEEE C37.118, specifies message size for different WAMPC applications, including types, use, contents, and data formats.

2.4 Summary

The smart grid depends on communication for monitoring and control, and therefore, it depends on a collection of different types of networks to communicate with a large number of advanced devices, applications, and systems on different layers. The communication layer has a crucial role in the overall smart grid environment to provide a two-way reliable, efficient, and secure data transmission for upper-layer applications. The network requirements, that is, data rate, latency, reliability, and security, vary greatly depending on the application.

This chapter provides a broad overview of available communication and networking technologies, both wired solutions (fiber optic, DSL, coaxial cable, and PLC) and wireless solutions (ZigBee, wireless mesh, Wi-Fi, Z-Wave, WiMAX, Cellular, LoRa, NB-IoT, LTE-M, and satellite). These technologies are discussed and compared in terms of pros/cons, data rates, and coverage ranges. In addition, this chapter also evaluates the network requirements – based on typical data message size, data sampling requirement, latency, and reliability for different smart grid applications in HAN/BAN/IAN, NAN/FAN, and WAN.

Considering network requirements, home, building, and industrial automation applications in HAN/BAN/IAN require low-cost and power consumption communication technologies; NAN applications, such as meter reading, DA, and DR, require high-reliability and low-latency communication technologies; and lastly, WAN applications require the highest data rate, lowest latency, and highest reliability and security among other smart grid applications.

References

1. R. Ma, H.H. Chen, Y.R. Huang, and W. Meng, "Smart Grid Communication: Its Challenges and Opportunities," *IEEE Transactions On Smart Grid,* vol. 4, no. 1, pp. 36-46, 2013.
2. NIST Framework and Roadmap for Smart Grid Interoperability Standards [Online]. Available: http://www.nist.gov/public_affairs/releases/upload/smartgrid_interoperability_final.pdf. Retrieved: July 2012.
3. M. Kuzlu, M. Pipattanasomporn. "Assessment of communication technologies and network requirements for different smart grid applications", 2013 IEEE PES Innovative Smart Grid Technologies Conference (ISGT), 2013
4. Passive Optical Network (PON) [Online]. Available: https://www.viavisolutions.com/en-us/passive-optical-network-pon. Retrieved: July 2020.
5. Wavelength-Division Multiplexing (WDM) [Online]. Available: https://www.fiberlabs.com/glossary/about-wdm/. Retrieved: July 2020.
6. The Differences Between SDH And SONET [Online]. Available: http://www.fowiki.com/b/the-di%EF%AC%80erences-between-sdh-and-sonet/. Retrieved: July 2020.
7. Digital Subscriber Line (DSL) [Online]. Available: https://www.cisco.com/c/en_uk/solutions/routing-switching/dsl.html Retrieved: July 2020.
8. A. Habib, H. Saiedian, "Channelized voice over digital subscriber line" IEEE Communications Magazine, vol. 40, no.10, pp. 94–100, October 2002.

9. What Is DOCSIS and Why Does It Matter? [Online]. Available: https://highspeedexperts.com/home-networking/what-is-docsis-and-why-does-it-matter/ Retrieved: July 2020.
10. N. Sagar, Powerline Communications Systems: Overview and Analysis. Diss. Rutgers University-Graduate School-New Brunswick, 2011.
11. M. Kuzlu, M. Pipattanasomporn and S. Rahman, "Review of communication technologies for smart homes/building applications," 2015 IEEE Innovative Smart Grid Technologies - Asia (ISGT ASIA), Bangkok, Thailand, 2015, pp. 1–6, https://doi.org/10.1109/ISGT-Asia.2015.7437036.
12. Wi-Fi 6 Explained: The Next Generation of Wi-Fi, [Online]. Available: https://www.techspot.com/article/1769-wi-fi-6-explained/ Retrieved: July 2020.
13. Discover Wi-Fi, [Online]. Available: https://www.wi-fi.org/discover-wi-fi Retrieved: July 2020.
14. Parvin, J. Rejina. "An Overview of Wireless Mesh Networks." In Wireless Mesh Networks-Security, Architectures and Protocols. IntechOpen, 2019.
15. Z-Wave, [Online]. Available https://z-wavealliance.org/about_z-wave_technology/ Retrieved: July 2020.
16. Z-Wave explained: What is Z-Wave and why is it important for your smart home?, [Online]. Available: https://www.the-ambient.com/guides/zwave-z-wave-smart-home-guide-281 Retrieved: July 2020.
17. How WiMAX Works [Online]. Available: https://computer.howstuffworks.com/wimax.htm Retrieved: July 2020.
18. What Is 5G? [Online]. Available: https://www.pcmag.com/news/what-is-5g Retrieved: July 2020.
19. Bingöl, E., Kuzlu, M. and Pipattanasompom, M., 2019, April. A LoRa-based smart streetlighting system for smart cities. In 2019 7th international Istanbul smart grids and cities congress and fair (ICSG) (pp. 66–70). IEEE.
20. K. Mekki, E. Bajic, F. Chaxel, F. Meyer, "A comparative study of LPWAN technologies for large-scale IoT deployment", ICT Express, 2018
21. Ericsson, Breaking new ground with NB-IoT in rural areas, [Online]. Available: https://www.ericsson.com/en/blog/2020/7/groundbreaking-nb-iot-in-rural-areas Retrieved: July 2020.
22. Lanner America, "LTE-M vs LoRa: Who Will Win The IoT Race?" [online]. Available: https://www.lanner-america.com/blog/lte-m-vs-lora-will-win-iot-race/. Retrieved: July 2020.
23. Dawaliby, S., Bradai, A. and Pousset, Y., 2016, October. In depth performance evaluation of LTE-M for M2M communications. In 2016 IEEE 12th International Conference on Wireless and Mobile Computing, Networking and Communications (WiMob) (pp. 1–8). IEEE.
24. Principles of Satellite Communications [Online]. Available: https://www.tutorialspoint.com/principles_of_communication/principles_of_satellite_communications.htm. Retrieved: July 2020.
25. Kuzlu, M., Pipattanasomporn, M. and Rahman, S., 2014. Communication network requirements for major smart grid applications in HAN, NAN and WAN. Computer Networks, 67, pp. 74–88.
26. Time of Use Energy Metering (TOU) [Online]. Available: https://www.accuenergy.com/application-solutions/time-of-use-metering/ Retrieved: July 2020.
27. How Does Prepaid Electricity Work? [Online]. Available: https://www.firstchoicepower.com/prepaid-learn/how-does-prepaid-electricity-work Retrieved: July 2020.
28. Demand Response: An Introduction, Overview of programs, technologies, and lessons learned, [Online]. Available: http://large.stanford.edu/courses/2014/ph240/lin2/docs/2440_doc_1.pdf Retrieved: July 2020.
29. Distribution Automation Feeder Automation Design Guide, [Online]. Available: https://www.cisco.com/c/en/us/td/docs/solutions/Verticals/Distributed-Automation/Feeder-Automation/DG/DA-FA-DG/DA-FA-DG.html. Retrieved: July 2020.

Chapter 3
Smart Grid Standards and Protocols

The term smart grid refers to a next-generation electrical grid that uses advanced information, communication, and computing technologies to operate more efficiently. These technologies also provide tremendous economic and environmental benefits to the electrical grid. With emerging smart grid technologies, the deployment of smart devices and applications has been significantly increased, leading to the development of a variety of smart grid-related standards and protocols. These standards and protocols have become a necessity for the seamless integration of and interoperability among smart devices and applications.

With emerging smart grid technologies, a variety of standards development organizations (SDOs) have been started to develop standards and protocols including all aspects of the smart grid, for example, enterprise and control center, wide area monitoring, substations automation, distributed energy resources (DERs), demand response (DR), metering, electric vehicles (EVs), cybersecurity, and home/building automation. These SDOs include the Institute of Electrical and Electronics Engineers (IEEE), International Electrotechnical Commission (IEC), International Telecommunication Union (ITU), Internet Engineering Task Force (IETF), National Institute of Standards and Technology (NIST), International Organization for Standardization (ISO), and others.

The objective of this chapter is to briefly review and discuss major standards, protocols, and challenges in the smart grid domain. This chapter first discusses major standards organizations, alliances and user groups, and open source groups dealing with smart grid standards in Section 3.1. Section 3.2 presents a comprehensive review of commonly used standards and protocols in the smart grid environment. Section 3.3 describes the challenges and future research directions that can foster the deployment of smart grid applications in the long run. Finally, Section 3.4 presents the summary.

© The Author(s), under exclusive license to Springer Nature
Switzerland AG 2021
U. Cali et al., *Digitalization of Power Markets and Systems Using Energy Informatics*, https://doi.org/10.1007/978-3-030-83301-5_3

3.1 Standards Organizations, Alliances/User Groups, and Open Source Groups Dealing with Smart Grid Standards

With a diverse set of smart grid technologies, power system operations are influenced by a large number of vendors, regulations, and standards. Standards are essential for a successful implementation of the smart grid because they ensure interoperability among smart devices from different vendors. This section gives an overview of key SDOs that are responsible for developing smart grid-related standards, as well as alliances and user groups. Alliances and user groups are formed by nonprofit and commercial entities and individuals to promote a particular technology that they believe has value in advancing the smart grid mission. Open source groups, a more recent development, are communities of practice consisting of companies and individual contributors who work on software systems for smart grid and publish the source code for review and modification. Selected SDOs, user groups, and open source groups dealing with smart grid standards, protocols, and software are summarized in Table 3.1.

3.2 Smart Grid Standards and Protocols

Figure 3.1 presents the smart grid standards and protocols. At the top standards are categorized in the following domains: (1) cybersecurity, (2) enterprise, control center, and wide area monitoring, (3) substation automation, (4) distributed generation and demand response, (5) metering, (6) electric vehicles, and (7) home/building automation. These standards and protocols are partly explained in [1]. The bottom of the figure contains a high-level view of the grid architecture.

3.2.1 Enterprise, Control Center, and Wide Area Monitoring

3.2.1.1 IEC 61970

IEC 61970, proposed by the International Electrotechnical Commission (IEC), is known as the Energy Management System Application Program Interface (EMS-API). IEC 61970 defines an information model with common objects for electric transmission systems and provides a semantic model, data exchange formats, integration processes, and an abstract communication interface, that is, application programming interface (API), for data exchange for energy management systems. It is a set of standards consisting of five parts: (1) guidelines and general requirements, (2) glossary, (3) common information model (CIM), (4) component interface specification (CIS), and (5) CIS technology mappings [2].

Table 3.1 An overview of organizations dealing with smart grid standards and protocols

Organizations	Notes
International standards organizations	
IEC (International Electro-technical Commission)	A nonprofit, nongovernmental standards organization, which encompasses a wide range of smart grid technologies, such as power generation, transmission, and distribution, substation automation, distributed energy resources
ISO (International Organization for Standardization)	An international standards body composed of representatives from a variety of national standards organizations. It fosters worldwide proprietary, industrial and commercial standards. ISO and IEC have set up a joint technical committee on the smart grid
IEEE (Institute of Electrical and Electronics Engineers)	A nonprofit professional association dedicated to enhancing technological innovation and excellence. It has more than 100 smart grid-related standards
ITU (International Telecommunication Union)	Responsible for issues that concern information and communication technologies. It has set up a focus group on Smart Grid to identify and analyze communication networking requirements and capabilities to support smart grid applications
ISA (International Society of Automation)	A nonprofit professional association, which produces the standard to be able to improve automation and control systems in terms of management, safety, and cybersecurity
IETF (Internet Engineering Task Force)	A nonprofit standards organization that develops standards for the Internet, including routing, protocols at the network layer and above, applications and services such as the Domain Name System (DNS) and security standards such as standardized cryptography suites
Standards organizations in the United States	
NIST (National Institute of Standard and Technology)	NIST is a part of the US Department of Commerce, which promotes and maintains measurement standards and has a leading role in coordinating smart grid standards
ANSI (American National Standards Institute)	ANSI is a private nonprofit organization, which provides a framework to improve the quality of secure and interoperable smart grid products
EPRI (Electric Power Research Institute)	A nonprofit organization funded by global energy companies, utilities, and technology providers. It focuses on addressing challenges in electric power systems, including reliability, efficiency, and the environment
NERC (North American Eclectic Reliability Corporation)	A nonprofit organization focusing on the reliability of the bulk power system in the US NREC develops and enforces reliability standards related to transmission system operation and assesses future adequacy of electricity
ASHRAE (American Society of Heating, Refrigeration, and Air-Conditioning Engineers)	A global technical society focusing on building systems, energy efficiency, indoor air quality, refrigeration, and sustainability within the industry. It proposes the "Facility Smart Grid Information Model" to facilitate the integration of objects and actions within the electrical infrastructure
Standards organizations in Europe	
ETSI (European Telecommunications Standards Institute)	A nonprofit organization that produces standards for information, communications, and networking technologies, including fixed, mobile, radio, converged, broadcast, Internet, etc.

(continued)

Table 3.1 (continued)

Organizations	Notes
CEN (European Committee for Standardization)	CEN is the European mirror body for ISO and a business facilitator in Europe. It plays a leading role with CENELEC in the European Smart Meter Coordination group on standards for the smart grid and electric vehicles
CENELEC (European Committee for Electrotechnical Standardization)	CENELEC is the European mirror body for IEC. It is a significant contributor to the smart grid standardization activities and helps prepare voluntary standards
Alliances and users groups	
HomePlug Alliance	HomePlug Alliance is a group of about 44 companies working together to develop technical specifications for powerline networking. It brings together individual researchers and technologists to provide an environment where powerline communications can be promoted
Z-Wave Alliance	Z-Wave Alliance consists of international industry leaders that develop and extend the Z-Wave device wireless networking standard for home area networks (HANs)
UtilityAMI	UtilityAMI is a group that provides a forum for utilities to define cybersecurity and interoperability guidelines for advanced metering infrastructure (AMI) and demand response (DR) infrastructures for utilities and energy service providers
OpenAMI	OpenAMI is a global task force from the electricity metering industry focus on establishing an open-standard communications architecture for metering and DR applications
OpenHAN	OpenHAN specifies standards for home networks to develop guidelines, use cases, and platform-independent requirements for utility AMI and home area networks (HANs)
AMI-SEC	AMI-SEC is a task force in charge of developing security guidelines and recommendations for AMI system elements. It provides a focal point for industry discussions on security issues related AMI
UCAIug	UCA International Users Group is a nonprofit corporation, including suppliers, electric utilities, consultants, etc., whose primary goal is to enable integration in the energy and utility industry through open standards
OpenSG	OpenSG is a user group consisting of utilities, vendors, consultants, governments, universities, etc. The main objective of OpenSG is to provide business and industry requirements for accelerating standards development and adoption
Open source groups	
LF Energy	LF (Linux Foundation) Energy is an open source initiative hosted at the Linux Foundation. LF Energy provides a neutral environment for collaboration on open source software for the digitalization of the power industry
Energy Web Foundation	Energy Web Foundation (EWF) is a global organization that focuses on building core infrastructure and shared software technology, promoting adoption of commercial solutions and fostering a practice community for blockchain and other decentralizing technologies applied to the energy industry

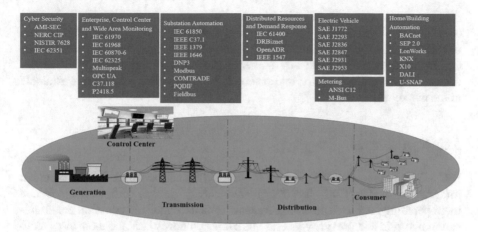

Fig. 3.1 Smart grid-related standards and protocols based on the power system domain

3.2.1.2   IEC 61968

IEC 61968 is proposed by the International Electrotechnical Commission (IEC) for information exchanges between electrical distribution systems, which is known as Application Integration at Electric Utilities-System Interfaces for Distribution Management. The main intention of the IEC 61968 standard is to be able to support the inter-application integration of a utility, which needs to collect data from a variety of distributed applications through different interfaces and middleware services. IEC 61968 defines interfaces and interface architecture to parts in the distribution management systems (DMS), and focuses on different application module interaction instead of system-level interaction.

IEC 61970/IEC 61968 standards together provide a common information model (CIM) in the smart grid environment. The primary goal of the CIM is to serve as a basis for the definition of interfaces in order to improve interoperability. IEC 61970 focuses on the transmission domain, while IEC 61968 focuses on the distribution domain. The CIM is defined and maintained in Unified Modeling Language (UML), and the specification and implementation artifacts are automatically derived from the UML model [3].

3.2.1.3   IEC 60870-6

IEC 60870-6, proposed by IEC, is known as Inter-Control Center Protocol (ICCP). IEC 60870-6 specifies systems used for telecontrol, that is, supervisory control and data acquisition (SCADA), referring to the connection of process stations spread out over a wide area to central control systems in power system automation applications. It provides a communication profile that enables communication through basic telecontrol messages between power system applications. ICCP provides a

data communication protocol for data transfer, monitoring, and control over wide
area networks between utility control centers. The ICCP protocol is a complete suite
of tools for SCADA, including management tools. It is based on client/server principles and does not provide any authentication or encryption. ICCP is widely
accepted by the utility industry, and several ICCP-compatible products are available
in the market [4].

3.2.1.4 IEC 62325

IEC 62325, proposed by IEC, is a set of standard frameworks to define the common
information model (CIM) for deregulated energy market communications and
interoperability. It also defines requirements for a decentralized common communication platform and provides general guidelines on how to use the e-business technologies, such as ebXML (e-business eXtensible Markup Language), and architecture
in energy markets. The intention of IEC 62325 is to support the integration of energy
market-based software solutions developed by independent vendors, into a market
management system, between market management systems and market participant
systems. The semantics of this message exchange is based on CIM [5].

3.2.1.5 MultiSpeak

MultiSpeak is an industry-wide interoperability standard, which is developed by the
National Rural Electric Cooperative Association (NRECA). MultiSpeak addresses
software integration and reliable data interoperability in the electric distribution
domain. It defines an information model and message structure in an XML schema
and a data exchange protocol based on web services, that is, Web Services
Description Language (WSDL). It also defines standardized interfaces among software applications commonly used by electric utilities to facilitate specific business
process steps in electric utilities. MultiSpeak provides schemas for different smart
grid applications, including meter reading, connect/disconnect, meter data management, outage detection, load management, SCADA, demand response, and distribution automation control [6]. The MultiSpeak community has grown significantly
and is used in more than 800 electric, water and gas utilities, universities, and
research groups worldwide [7].

3.2.1.6 IEEE C37.118

IEEE C37.118 is an IEEE standard, which was developed for synchronized phasors,
that is, synchrophasors, in the smart grid applications. It defines synchrophasor frequency and rate of change of frequency measurement, as well as the communication
protocol for data exchange. It also defines the phasor measurement unit (PMU),
which is a single-power system device measuring the synchronized voltage and

current phasor in a power system. IEEE C37.118 also specifies phasor measurements and test procedures to ensure that measurements follow the proposed format [8]. The protocol supports Ethernet, IP, or Fieldbus technologies for data transmission.

3.2.1.7   P2418.5

P2418.5 is an open, interoperable reference model framework, which was developed for blockchain-based applications in the smart grid environment. It was developed by an IEEE P2418.5 working group under the IEEE Standards Association (IEEE-SA). The intention of the proposed model is to facilitate the integration of blockchain applications, such as peer-to-peer energy trading, and EV charging into the energy sector. It provides a guideline for blockchain use cases in the energy sector, including the electrical power industry and the oil and gas industry. It also defines the standardized reference architecture, interoperability, terminology, and system interfaces for blockchain use cases through an open protocol and technology agnostic framework. Lastly, it evaluates and provides guidelines on scalability, performance, security, and interoperability through the evaluation of blockchain components, including consensus algorithms and smart contracts [9].

3.2.2   Substation Automation

3.2.2.1   IEC 61850

The IEC 61850 standard provides interoperability among intelligent electronic devices (IEDs) for protection, monitoring, control, and automation in substations. The main objective of this standard is to ensure compatibility with the CIM for power system applications. To distinguish the power system applications and prioritize traffic flows, IEC 61850 defines five types of communication services: (1) abstract communication service interface (ACSI), (2) generic object-oriented substation event (GOOSE), (3) generic substation status event (GSSE), (4) sampled measured value multicast (SMV), and (5) time synchronization (TS). It is based on the Ethernet and IP standards and enables seamless data communications and information exchange among devices in a distribution network. IEC 61850 is used by a number of manufacturers on their power system products and equipment, including distribution automation nodes, grid measurement, and diagnostics devices [10].

3.2.2.2   IEEE C37.1

IEEE C37.1 is an IEEE Standard, which was developed for SCADA and automation systems. IEEE C37.1 provides the basis for a definition, specification, performance analysis, and application of SCADA and automation systems in power electric substations. It also defines system architectures and functions in a substation, from the protocol selection to human–machine interfaces to implementation issues. Network performance requirements are also defined by IEEE C37.1 [11].

3.2.2.3 IEEE 1379

IEEE 1379 is an IEEE standard, which was developed for communications and interoperation of IEDs and remote terminal units (RTUs) in an electric utility sub-station. IEEE 1379 provides implementation guidelines and a recommended prac-tice, including a basic set of data objects for adding data elements and message structures. IEEE 1379 helps also reduce the time and cost of deploying IEDs and RTUs by facilitating the integration of devices through standardized interfaces [12].

3.2.2.4 IEEE 1646

IEEE 1646 is an IEEE standard that specifies performance requirements for com-munication delivery times in electric power substation automation. IEEE 1646 defines the time required for applications running at two end systems in a network segment to communicate, including both processing and transmission delays for data exchange among power system equipment. Substation communications are classified into different categories defined in IEEE 1646 in terms of communication delay requirements. Each category has different network requirements, for example, a system protection message is required to be transmitted within four milliseconds while an operation maintenance message within 1 s [13].

3.2.2.5 DNP3

Distributed Network Protocol (DNP3) is a set of communications protocols used between devices in automation systems by electric and water utilities. DNP3 plays a key role in SCADA systems, where it is primarily used for communication proto-cols among SCADA Master Stations (Control Centers), RTUs, and IEDs. DNP3-enabled devices can exchange status and control information to automate substation management. DNP3 uses the IP suite to transport data messages for equipment monitoring and control in an electric power substation. In addition, it provides the possibility for file transfer and time synchronization [14].

3.2.2.6 Modbus

Modbus is the most popular serial communication protocol and is widely used in a large number of applications, such as industrial, buildings, energy, etc. It was origi-nally developed by Modicon in 1979 for use with programmable logic controllers

(PLCs). Modbus offers a common format for the layout and content of message fields. A Modbus packet consists of a messaging structure designed to establish a controller–worker, that is, or client–server, communication architecture to connect a supervisory computer with remote terminal units (RTUs) in a SCADA system. It supports both serial and Ethernet protocols. There were two transmission modes in the original version of Modbus, that is, ASCII and RTU. More recently, Modbus/TCP has been developed, allowing data transmission over TCP/IP-based networks. It allows for communications among a number of devices (up to 247) in the same network [15].

3.2.2.7   Open Platform Communications United Architecture (OPC UA)

Open Platform Communications United Architecture (OPC UA) is developed by the OPC Foundation and standardized as IEC 62541. OPC UA is a data exchange standard defining the communication infrastructure and information model for industrial automation applications. It can run directly on intelligent devices, systems, and machines having real-time-capable operation systems. Data can be encoded using XML for interoperability of web service applications, that is, HTTP/SOAP-based. OPC UA communication is based on a client–server model. For adopting OPC UA to the smart grid applications, important data structures, for example, CIM and IEC 61850, have been identified to be integrated into OPC UA communications [16].

3.2.2.8   IEEE C37.111 (COMTRADE)

COMTRADE (Common Format for Transient Data Exchange) is standardized by the IEEE Power System Relaying Committee (PSRC) as C37.111, and is a file format for storing oscillography and status information regarding transient power system disturbances. COMTRADE files are generated by IEDs, which record electrical characteristics of a power system (e.g., current, voltage, power, and frequency) by digitally sampling these measurements at high speed and high data resolution. Collected files in COMTRADE format from a variety of power electric substations are used to analyze large-scale power disturbance events. COMTRADE facilitates understanding the cause of disturbance events and planning possible mitigation strategies for future events as well as predictive maintenance [17].

3.2.2.9   IEEE 1159.3 (PQDIF)

PQDIF (Power Quality Data Interchange Format) was initiated by IEEE and EPRI to unify data formats from simulations, measurements, and analysis tools for power quality engineers, and finally standardized by IEEE as 1159.3. It is a recommended practice for a file format, similar to COMTRADE, for exchanging power quality-related measurement and simulation data, including voltage, current, power, and energy measurements, in a vendor-independent manner. PQDIF is used mostly to transport the power quality data instead of transient disturbance data [18].

3.2.2.10   IEC 61158 (Fieldbus)

Fieldbus is a set of industrial computer network protocols standardized by IEC as 61158, which is used for real-time distributed control. The Fieldbus protocol connects programmable logic controllers (PLCs) components, such as sensors, lamps, actuators, electric motors, switches, valves, and others, at the end of the network. It allows connecting multiple field distribution devices to a single point then to the controller, that is, PLC, in a variety of network structures, such as a star, ring, branch, and daisy-chain. With the advancement of the Fieldbus protocol, field distribution devices can support control schemes, such as proportional integral derivative (PID) control, on the device side, rather than on the controller, to do the processing through some Fieldbus protocols [19]. There are many Fieldbus technologies in the market, such as EtherNet/IP DeviceNet, ControlNet, CompoNet, HART, PROFIBUS, PROFINET, CANopen, Modbus TPC, IO-Link, and others [20].

3.2.3   Distributed Resources and Demand Response

3.2.3.1   IEC 61400

IEC 61400 is an IEC standard that is proposed for uniform data exchange, especially in monitoring and control of wind power plants. The IEC 61400 standard specifies a vendor-independent format for communication between a utility control center and wind power plants. It also provides a set of design requirements for many wind power plant components to ensure that wind turbines are appropriately designed and installed in a way that prevents damage during operation. These design requirements documented in the IEC 61400 standard cover the following parts: (1) general design requirements, (2) design requirements for small wind turbines, (3) acoustic noise measurement techniques, (4) wind turbine power performance testing, (5) measurement of mechanical loads, (6) declaration of apparent sound power level and tonality values, (7) measurement and assessment of power quality characteristics of grid-connected wind turbines, (8) full-scale structural testing of rotor blades, and (9) lightning protection [21].

3.2.3.2   IEEE 1547

IEEE 1547 is a set of IEEE standards, which defines requirements for interconnecting and interoperability between distributed energy resources (DERs) and utility electric power systems (EPSs) to maintain system reliability in the smart grid environment. It provides requirements for performance, operation, testing, and safety. The intention of IEEE 1547 is to address the physical and electrical interconnection and interoperability issues regarding DERs with EPSs and provides information modeling, use-case approaches, and an information exchange template. This set of standards covers the following parts: (1) IEEE 1547.1 – Conformance tests procedures for equipment interconnecting distributed resources with EPS, (2) IEEE 1547.2 – Application guide for interconnecting distributed resources with EPS, (3) IEEE 1547.3 – Guide for cybersecurity of distributed resources with EPS, (4) IEEE 1547.4 – Guide for design, operation, and integration of distributed resource island systems with EPS, (5) IEEE 1547.6 – Recommended practice for interconnecting DERs with EPS distribution secondary network, and (6) IEEE 1547.7 – Guide to conducting distribution impact studies for distributed resource interconnection [22].

3.2.3.3 DRBizNet

Demand Response Business Network (DRBizNet) is a flexible demand response management platform to simplify DR applications and enable efficient DR programs with distributed business process integration technologies in the smart grid environment. DRBizNet provides a service-oriented architecture through a standardized web services interface. Market operators, utilities, and aggregators can manage DR programs efficiently, reliably, and securely through the DRBizNet platform. The platform provides alarm and automatic notifications to customers, aggregators, and utilities. It also triggers any type of intelligent devices, such as load controllers, programmable thermostats, and energy management systems [23].

3.2.3.4 OpenADR

Open Automated Demand Response (OpenADR) proposed by OpenADR Alliance is an open communication protocol to standardize, automate, and simplify the DR applications. It provides data exchange between utilities or electricity service providers and their customers, and a communication interface to electricity providers to be able to send a DR signal directly to existing customers using a common language (SOAP and XML) through web services. OpenADR also specifies the message content and format used during DR and DER control events. This message content includes a variety of information, such as event name/identification/status, reliability and emergency signals, operating mode, renewable generation status, and price signals. The typical DR signal is sent to turn OFF the selected load(s) during a DR period, that is, high electricity demand [24].

3.2.4 Metering

3.2.4.1 ANSI C12

ANSI C12 is a standard suite consisting of ANSI C12.18, 12.19, 12.20, 12.21, 12.22 for electricity meters, and their accuracy and performance and is mostly used in the USA. It defines requirements and guidance for electricity metering, watt-hour meter sockets, end-device data tables, accuracy classes, as well as the interface to data communication networks [25]. In addition, the suite defines the performance criteria for thermal demand meters, mechanical demand registers, and phase-shifting devices used in smart metering [26].

3.2.4.2 M-Bus

Meter Bus (M-Bus) is a European standard, that is, EN 13757, for remote meter reading, such as electricity and gas. M-Bus is designed for low-cost, battery-powered devices. In addition to metering, M-Bus can be used for other applications, such as alarm systems, lighting, and heating control [27]. An M-Bus metering infrastructure has been specified in Dutch Smart Meter Requirements [28] as the means of communications between a utility and utility meters with improved security (AES instead of DES).

3.2.4.3 Electric Vehicles (EVs)

Electric vehicle (EVs) standards are proposed by the Society of Automotive Engineers (SAE) [29]. The intention of the proposed EV standards is to facilitate the widespread acceptance and deployment of EV and related applications.

3.2.4.4 SAE J1772

SAE J1771 is a North American standard known as either "Electric Vehicle and Plug-in Hybrid Electric Vehicle (PHEV) Conductive Charge Coupler" or "J Plug." It specifies general physical, electrical, functional, and performance requirements to facilitate conductive charging of EV/PHEV in North America and, in addition, different charging methods and connector requirements, such as level 1, level 2, and DC charger.

3.2.4.5 SAE J2293

SAE J2293 is a North American standard known as the "Energy Transfer System for Electric Vehicles". SAE J2293 specifies requirements for EVs and the off-board electric vehicle supply equipment (EVSE). These requirements include system, communication, and network architecture as well as functional interoperability requirements to be able to transfer electrical energy to an EV from an electric utility. SAE J2293 defines the energy form, that is, AC and DC, transferred between the EV and the EVSE.

3.2.4.6 SAE J2836

SAE J2836 defines EV-related communication use cases. They include communication between plug-in vehicles and the utility grid for reverse power flow, communication between plug-in vehicles and their customers, wireless charging communication between plug-in electric vehicles, and the utility grid. Additionally, it also defines diagnostic communication use cases for plug-in vehicles (PEVs) [30].

3.2.4.7 SAE J2847

SAE J2847 defines specifications and requirements for communication protocols and messaging in a variety of use cases. These use cases cover the communication between a PEV and an off-board DC charger, between PEV and the utility grid for energy transfer, between PEV and their customers, and between wireless charged EVs and wireless EV chargers. In addition, SAE J2847 defines the diagnostic communication for plug-in vehicles between a PEV and EVSE.

3.2.4.8 SAE J2931

SAE J2931 is a set of standards that defines communications and security for PEVs. SAE J2931 specifies the requirements for digital communication between PEVs, wireless charging PEVs, customers, energy service providers (ESP), and HAN, as well as the broadband PLC communication for PEVs. Additionally, it also defines the security requirements for PEV communications.

3.2.4.9 SAE J2953

SAE J2953 standard addresses EVSE-PEV interoperability, specifically how a PEV and an EVSE pair can interoperate. SAE J2953 defines test procedures to ensure EVSE-PEV interoperability from different vendors. It defines three levels of interoperability testing: (1) mechanical interoperability, charge, and safety feature functionality; (2) indefinite grid and dynamic grid events; and (3) ampacity control, scheduled and staggered scheduled charge, how to interrupt and resume charging.

3.2.5 Cybersecurity

There are a variety of standards and use cases that are applicable to cybersecurity for smart grid applications, many of which are derived from the broader set of Internet security communication standards. This section only addresses the major cybersecurity standards and requirements specifically related to the smart grid. A list of other cybersecurity-related standards can be obtained from the UCA International Users Group (UCAIug) [31].

3.2.5.1   AMI System Security Requirements (AMI-SEC)

AMI-SEC was developed by the AMI-SEC Task Force to provide a set of cybersecurity requirements for advanced metering infrastructure (AMI) [32]. These security requirements aim to serve as a guide for utilities and vendors. AMI-SEC covers modules of the entire AMI system, including communications and networking devices, forecasting systems, the meters, and the meter and data management system to the HAN interface of the smart meter. According to AMI-SEC, AMI system security requirements are categorized into three areas: (1) primary security services (confidentiality and privacy, integrity, availability, identification, authentication and authorization, non-repudiation and accounting), (2) supporting security services (anomaly detection, cryptographic, notification and signaling, resource management and trust and certificate), and (3) assurance services (development/organizational/handling/operating rigor, accountability, and access control) [33].

3.2.5.2   NERC CIP

NERC CIP (North American Electric Reliability Corporation Critical Infrastructure Protection) standard defines a set of requirements to secure assets that are critical for operating bulk electric power systems in North America and is primarily of interest to transmission system operators. NERC CIP covers a variety of topics [34], such as identifying and categorizing assets, personnel and training in cybersecurity, security management controls, electronic security perimeter(s), cybersecurity incident reporting and response planning, vulnerability management, information protection, recovery plans for critical cyber assets, as well as security plans to limit physical and electronic access.

3.2.5.3   NISTIR 7628

NISTIR 7628 provides guidelines for "Smart Grid Cybersecurity." It consists of a three-volume report: (1) smart grid cybersecurity strategy, architecture, and high-level requirements; (2) privacy and the smart grid; and (3) supportive analyses and references [35]. These reports introduce a comprehensive framework, which can help organizations develop effective smart grid cybersecurity strategies, as well as provide guidance to assess cybersecurity risks and help identify appropriate cybersecurity requirements. "Guidelines for Smart Grid Cybersecurity" was initially issued in November 2009 as a companion developed by the Cyber Security Working Group (CSWG) under the Smart Grid Interoperability Panel (SGIP), and then launched by the National Institute of Standards and Technology (NIST) [36].

3.2.5.4   IEC 62351

IEC 62351 is a set of standards that defines cybersecurity requirements for power systems management and securing power system communications. IEC 62351 standard group consists of the following parts: (1) IEC 62351-1 – Introduction to the IEC 62351 standards, (2) IEC 62351-2 – Terms and acronyms used in IEC 62351 standards, (3) IEC 62351-3 – Security for any Internet communication profile including TCP/IP, (4) IEC 62351-4 – Security for any telecommunication profile including Multimedia Message Specification (MMS), (5) IEC 62351-5 – Security for IEC 60870-5, (6) IEC 62351-6 – Security for IEC 61850 profiles, (7) IEC 62351-7 – Security for network and system management, (8) IEC 62351-8 – Access control mechanisms, (9) IEC 62351-9 – Key management, (10) IEC 62351-10 – Security architecture, and (11) Security for XML files [37].

3.2.6   Home/Building Automation

3.2.6.1   BACnet

The Building Automation and Control Network (BACnet) is a communication protocol developed by ASHRAE for building automation and control networks. It supports the interoperability of building automation, management, and control systems from multiple vendors. BACnet defines an information model with object types in building automation applications, such as heating, ventilation, and air conditioning (HVAC) control and lighting control. BACnet supports security for client–server communications. Authentication is performed by the handshaking process, and messages are encrypted using a session key [38].

3.2.6.2   Smart Energy Profile (SEP) 2.0

Smart Energy Profile (SEP) 2.0 standard is designed to serve as an interoperable protocol to connect energy devices in a home environment with the grid. SEP 2.0 was developed by some technological user groups, such as Wi-Fi Alliance, ZigBee Alliance, HomePlug Alliance, and HomeGrid Alliance. SEP was originally proposed as ZigBee SE 1.x and was very simple and limited in terms of the services and security features it supported. SEP 2.0 is an enhancement of the ZigBee SE 1.x with added services and security for PEV charging, installation, configuration, prepay services, metering, pricing, user information and messaging, load control, demand response events, etc. [39]. SEP 2.0 uses a representational state transfer (REST) architecture widely for offering web service APIs over the Hypertext Transfer Protocol (HTTP).

3.2.6.3   LonWorks

LonWorks is one of the most widely deployed open control standards worldwide and is built on a protocol created by Echelon Corporation for networking devices. Today, many LonMark-certified products are commercially available for home and building automation applications. The LonWorks protocol is standardized as ISO/IEC 14908 (worldwide), EN 14908 (Europe), and ANSI/CEA-709/852 (US). The protocol is used for monitoring and controlling HVAC, lighting, and metering, etc., in a building. The LonWorks community is also expanding its scope to cover entire building control systems connecting to the Internet [40].

3.2.6.4   KNX

KNX is a communication protocol for home and building automation. The KNX protocol is widely used worldwide under different standards, that is, ISO/IEC 14543-3 (international standard), CSA-ISO/IEC 14543-3 (Canadian standard), CENELEC EN 50090 and CEN EN 13321-1 (European standard), and GB/T 20965 (Chinese standard). KNX is administered by the KNX association. It was designed as an independent hardware platform that can be implemented by devices having a modest size. Most forms of KNX installation run over a twisted pair of a physical network. KNX can be used for a variety of functions and applications in home and building control, for example, HVAC, lighting, appliances, alarm systems, sensors, energy management, metering, etc. [41].

3.2.6.5   X10

X10 is a protocol for signaling and control communication between electronic devices used for home automation that runs over the home power line wiring. A wireless radio-based communication protocol is also defined for X10. It was developed by Pico Electronics in 1975 and was the first general-purpose home networking technology to allow remote control of smart home devices and appliances. Although a number of higher bandwidth alternatives are available in the market, X10 is still popular in home automation applications today [42].

3.2.6.6   DALI

Digital Addressable Lighting Interface (DALI), also specified in IEC 60929, is a data communication and transport mechanism cooperatively developed by several lighting equipment vendors. DALI defines a standardized interface for lighting control that enables equipment from different vendors to interconnect, interoperate, and communicate with the building automation and control system. It provides simplified installation and communication, yet maximum control and flexibility, and allows for a maximum of 64 fittings on a single network segment, with the network broken into 16 different possible segments [43].

3.2.6.7 U-SNAP

U-SNAP (Utility Smart Network Access Port) was developed to address the lack of standardization in HANs. U-SNAP standardizes a connector and serial interface by specifying the hardware interface, physical dimensions, data transfer, message contents, and protocol for HAN devices. U-SNAP standard facilitates the connection of HAN devices that operate on a number of communication protocols, for example, Wi-Fi, ZigBee, Z-Wave, etc., to smart meters [44].

3.3 Challenges

Many smart technologies have been deployed to enable two-way communication with the aim of allowing finer-grained control over electrical transmission and distribution equipment and end loads. One major goal is to enable interoperability among smart grid applications deployed on various devices and platforms. While significant progress is being made toward the development and implementation of a standards-compliant smart grid, there are still many challenges that need to be addressed [45].

1. *The major challenge is device, application, and system interoperability.* The interoperability is defined as the ability of different devices, applications, and systems to be able to exchange information and work together in a coordinated manner. Interoperability is the key to enable integration, effective cooperation, and two-way communications among many interconnected elements. An urgent need exists to develop interoperability standards that allow utilities to deploy smart components from any vendor, knowing that they will work with existing power system components at every level. As the leading smart grid SDO in the US, NIST has established the following three-phase approach to identify missing smart grid standards [46]:

   - Phase 1 engages actors in a public process to identify applicable smart grid-related standards and gaps in currently available standards and define priorities for new standardization activities.
   - Phase 2 establishes a private–public partnership and shapes a smart grid interoperability panel to drive longer-term progress.
   - Phase 3 implements a framework for testing and certification of how standards are implemented in smart grid devices, applications, systems, and processes.

2. *The second is a lack of awareness.* There are many smart grid standards, and best practices are available and can be readily applied to facilitate smart grid applications. However, their adoption seems a big issue due to the lack of awareness of those standards by the organizations involved in designing and deploying smart grid systems and the lack of clear best practices and guidelines for applying them.
3. *The third is the technical* challenges. The electric power grid consists of many components operating dependently. As the implementation of smart devices has already begun, one issue is that smart devices integrated into a smart grid have a lower lifespan than traditional network assets. That is, consumer electronics and telecommunications may have a lifespan of 3–5 years, while the lifespan of traditional electrical system distribution equipment such as lines, cables, and transformers is 40 plus years. One reason for such a shorter lifetime is the accelerated pace of development in information and communication technology compared with traditional electrical system technology. The other aspect is the fact that each smart device added to the smart grid environment serves as an electrotechnical device and, at the same time, an intelligent node. As a result, existing standards and protocols need to address both aspects of concurrently.
4. *The fourth is complexity.* The smart grid is a very complex system, consisting of many subsystems and parts. Many smart grid standards and protocols by different SDOs are developed to be used in smart grid projects. The standards from different SDOs are supposed to be compatible because they deal with components at many different levels and from different viewpoints, but in fact, there are areas of incompatibility. This is because each SDO develops standards and protocols from a different perspective, with different interests, needs, and criteria.

3.4 Summary

Smart grid standards and protocols have been proposed and developed by many organizations and user groups to guide the improvement of the smart grid. This chapter reviewed commonly used standards and protocols for all kinds of applications in the smart grid domain, such as enterprise and control center, wide area monitoring, substations automation, distributed resources, DR, metering, EVs, cybersecurity to home and building automation. Most commonly used smart grid standards and protocols are based on the specific domain: (1) Enterprise, Control Center And Wide-Area Monitoring – IEC 61970, IEC 61968, IEC 60870-6, IEC 62325, Multispeak, OPC UA, C37.118, P2418.5; (2) Substations Automation – IEC 61850, IEEE C37.1/1379/1646, DNP3, Modbus, COMTRADE, PQDIF, Fieldbus; (3) Distributed Resources and Demand Response – IEC 61400, DRBizNet, OpenADR, IEEE 1547; (4) Metering-ANSI C12, M-Bus; (5) EVs-SAE J1772/ J2293/J2836/J2847/J2931/J2953; (6) Cybersecurity-IEC 62351; and (7) Home and Building Automation – BACnet, SEP 2.0, LonWorks, KNX, X10, DALI, U-SNAP. The challenges in smart grid standards are discussed in the chapter.

References

1. M. Kuzlu, M. Pipattanasompom, S. Rahman. "A comprehensive review of smart grid related standards and protocols", 2017 5th International Istanbul Smart Grid and Cities Congress and Fair (ICSG), 2017.
2. Santodomingo R, Uslar M, Specht M, Rohjans S, Taylor G, Pantea S, Bradley M, McMorran A. IEC 61970 for energy management system integration. Smart Grid Handbook. 2016 Jul 29:1-29.
3. IEC CIM architecture for Smart Grid to achieve interoperability, [Online]. Available: http://www.gridwiseac.org/pdfs/forum_papers11/ambrosio_paper_gi11.pdf. Retrieved: July 2019.
4. IEC 60870-6 (ICCP), [Online]. Available: http://xanthus-consulting.com/IntelliGrid_Architecture/New_Technologies/Tech_IEC_60870-6_(ICCP).htm. Retrieved: July 2019.
5. Sabari Chandramohan, L.; Ravikumar, G.; Doolla, S.; Khaparde, S.A., "Business Process Model for Deriving CIM Profile: A Case Study for Indian Utility," Power Systems, IEEE Transactions on, vol.30, no.1, pp.132,141, Jan. 2015, DOI: https://doi.org/10.1109/TPWRS.2014.2324826.
6. MultiSpeak, [Online]. Available: http://www.multispeak.org/Pages/default.aspx. Retrieved: July 2019.
7. Multispeak - An interoperability standard for data systems integration, [Online]. Available: https://www.cooperative.com/programs-services/bts/Pages/MultiSpeak.aspx. Retrieved: Jan 2019.
8. C37.118.1-2011 - IEEE Standard for Synchrophasor Measurements for Power Systems, [Online]. Available: https://standards.ieee.org/findstds/standard/C37.118.1-2011.html. Retrieved: Feb 2019.
9. U. Cali, C. Lima, X. Li and Y. Ogushi, "DLT / Blockchain in Transactive Energy Use Cases Segmentation and Standardization Framework," 2019 IEEE PES Transactive Energy Systems Conference (TESC), 2019, pp. 1-5, doi: https://doi.org/10.1109/TESC.2019.8843372.
10. Mackiewicz, R.E., "Overview of IEC 61850 and benefits," Power Engineering Society General Meeting, 2006. IEEE, vol., no., pp.8 pp., 0-0 0, https://doi.org/10.1109/PES.2006.1709546.
11. IEEE Standard for SCADA and Automation Systems - Redline," IEEE Std C37.1-2007 (Revision of IEEE Std C37.1-1994) - Redline, vol., no., pp.1,200, May 8, 2008.
12. 1379-2000 - IEEE Recommended Practice for Data Communications Between Remote Terminal Units and Intelligent Electronic Devices in a Substation, [Online]. Available: https://standards.ieee.org/findstds/standard/1379-2000.html. Retrieved: Dec 2019.
13. 1646-2004 - IEEE Standard Communication Delivery Time Performance Requirements for Electric Power Substation Automation, [Online]. Available: http://standards.ieee.org/findstds/standard/1646-2004.html. Retrieved: Feb 2019.
14. The Distributed Network Protocol, [Online]. Available: https://www.dnp.org/default.aspx.
15. Modbus, [Online]. Available: http://www.modbus.org/. Retrieved: March 2020.
16. OPC Unified Architecture Specification, [Online]. Available: https://scadahacker.com/library/Documents/ICS_Protocols/OPCF%20-%20OPC-UA%20Part%201%20-%20Overview%20and%20Concepts%201.02%20Specification.pdf. Retrieved: Feb 2019.
17. IEEE Standard Common Format for Transient Data Exchange (COMTRADE) for Power Systems," IEEE Std C37.111-1991, vol., no., pp.1, 1991, https://doi.org/10.1109/IEEESTD.1991.101025.
18. IEEE Std. 1159-3 PQDIF, [Online]. Available: http://www.pqview.net/pqdif.asp. Retrieved: Feb 2019.
19. Fieldbus Foundation, [Online]. Available: http://www.fieldbus.org/. Retrieved: Feb 2019.
20. Network and Field Bus Technology, [Online]. Available: https://www.motioncontrolonline.org/products/network-field-bus.cfm#:~:text=The%20different%20types%20of%20fieldbus,Interface%3B%20and %20IO%2DLink. Retrieved: Feb 2019.
21. EIC 61400-1, [Online]. Available: https://webstore.iec.ch/preview/info_iec61400-1%7Bed3.0%7Den.pdf. Retrieved: Feb 2019.

22. IEEE Standards Coordinating Committee 21 (SCC21), [Online]. Available: http://grouper.
    ieee.org/groups/scc21/1547_series/1547_series_index.html. Retrieved: Feb 2020.
23. Designing a Demand Response Business Network for California, [Online]. Available: http://
    eetd.lbl.gov/news/events/2005/03/08/designing-a-demand-response-business-network-for-
    california. Retrieved: Feb 2020.
24. The OpenADR Alliance, [Online]. Available: http://www.openadr.org/. Retrieved: Feb 2019.
25. Snyder, A.F.; Stuber, M.T.G., "The ANSI C12 protocol suite - updated and now with net-
    work capabilities," Power Systems Conference: Advanced Metering, Protection, Control,
    Communication, and Distributed Resources, 2007. PSC 2007, vol., no., pp.117,122, 13-16
    March 2007, https://doi.org/10.1109/PSAMP.2007.4740906.
26. ANSI C12, [Online]. Available: https://webstore.ansi.org/industry/smartgrid/ansi-c12.
    Retrieved: Feb 2019.
27. The M-Bus: An Overview, [Online]. Available: http://www.m-bus.com/. Retrieved: Feb 2019.
28. P1 Companion Standard, [Online]. Available: http://files.domoticaforum.eu/uploads/
    Smartmetering/DSMR%20v4.0%20final%20P1.pdf. Retrieved: Feb 2019.
29. SAE Standards, [Online]. Available: http://standards.sae.org/. Retrieved: Feb 2019.
30. J2836/3_201301, [Online]. Available: https://www.sae.org/standards/content/j2836/3_201301.
    Retrieved: Feb 2019.
31. Cleveland, Francis. "List of Cybersecurity for Smart Grid Standards." 2013. http://iectc57.
    ucaiug.org/wg15public/Public%20Documents/List%20of%20Smart%20Grid%20Standa
    rds%20with%20Cybersecurity.pdf. Retrieved: Feb 2019.
32. Introduction to NISTIR 7628 Guidelines for Smart Grid Cyber Security, [Online]. Available:
    http://www.nist.gov/smartgrid/upload/nistir-7628_total.pdf. Retrieved: Feb 2019.
33. Security profile for advanced metering infrastructure, [Online]. Available: http://osgug.ucaiug.
    org/utilisec/amisec/Shared%20Documents/AMI%20Security%20Profile% 20 (ASAP-SG)/
    AMI%20Security%20Profile%20-%20v2_1.pdf. Retrieved: Feb 2019.
34. CIP Standards, http://www.nerc.com/pa/Stand/Pages/CIPStandards.aspx. Retrieved: Feb 2019.
35. Introduction to NISTIR 7628 Guidelines for Smart Grid Cyber Security, [Online]. Available:
    http://www.nist.gov/smartgrid/upload/nistir-7628_total.pdf. Retrieved: Feb 2019.
36. Guidelines for Smart Grid Cybersecurity Volume 1 - Smart Grid Cybersecurity Strategy,
    Architecture, and High-Level Requirements, [Online]. Available: http://nvlpubs.nist.gov/nist-
    pubs/ir/2014/NIST.IR.7628r1.pdf. Retrieved: Feb 2019.
37. EC 62351 Security Standards for the Power System Information Infrastructure, [Online].
    Available: http://iectc57.ucaiug.org/wg15public/Public%20Documents/White%20Paper%20
    on%20Security%20Standards%20in%20IEC%20TC57.pdf. Retrieved: Feb 2019.
38. BACnet, [Online]. Available: http://www.bacnet.org/. Retrieved: Feb 2019.
39. The Consortium for SEP 2 Interoperability, [Online]. Available: http://www.csep.org/.
    Retrieved: Feb 2019.
40. LonMark International, [Online]. Available: http://www.lonmark.org/. Retrieved: Feb 2019.
41. KNX Association, [Online]. Available: http://www.knx.org/. Retrieved: Feb 2019.
42. X10 devices and standards, [Online]. Available: http://www.x10.com. Retrieved: Feb 2019.
43. Communication in building automation, [Online]. Available: http://www.downloads.siemens.
    com/download-center/Download.aspx?pos=download&fct=getasset&id1=A6V10209534.
    Retrieved: Feb 2019.
44. USNAP Alliance, [Online]. Available: http://www.usnap.org/. Retrieved: Feb 2019.
45. Schneiderman, R., Smart Grid Takes on Critical Standards Challenges, Wiley-IEEE Standards
    Association, 2015.
46. Technology, Measurement, And Standards Challenges For The Smart Grid, [Online]. Available:
    http://www.nist.gov/smartgrid/upload/Final-Version-22-Mar-2013-Smart-Grid-Workshop-
    Summary-Report.pdf. Retrieved: Feb 2020.

Chapter 4
Introduction to Security for Smart Grid Systems

Digitized power systems offer high-value targets for attackers that are intent on disruption or illicit financial gain. Preventing attacks and mitigating them when they happen is an important part of ensuring the availability and quality of service in a smart grid power system, as is protecting the confidentiality of the customers' personal protected information. Many methodologies and technologies that are available from other Internet-connected systems can be brought to bear on the prevention, detection, and mitigation of attacks in smart grid systems.

In this chapter, we discuss the basics of security for smart grid systems. We begin in Section 4.1 with a description of how to conduct threat modeling and discuss the architectural principles that result from a consideration of how to design systems to mitigate threats. Section 4.2 reviews the collection of security technologies developed for the Internet, which have been applied to the problem of security for smart grids. In Section 4.3, we describe how the technologies in Section 4.2 can be used to address the threats discovered during a threat modeling exercise. These designs then become input to the security architecture, which in turn becomes part of the overall system architecture. In Section 4.4, we describe what role processes and people play in maintaining security. Finally, Section 4.5 pulls together considerations from the previous sections into an example smart grid security reference architecture.

4.1 Threat Modeling and Architectural Principals

The first step in designing a security architecture for a smart grid system is to conduct systematic modeling of possible threats [1]. A threat is a promise of future harm, and a vital part of securing smart grid systems is modeling threats. Threat modeling starts during the design phase of the system and involves asking the following questions:

© The Author(s), under exclusive license to Springer Nature
Switzerland AG 2021
U. Cali et al., *Digitalization of Power Markets and Systems Using Energy Informatics*, https://doi.org/10.1007/978-3-030-83301-5_4

- What are the services of the smart grid system being implemented?
- What aspect of these services offers high-value targets for attackers and openings for attacks, and what implementation and deployment techniques, omissions, or errors could leave the system vulnerable?
- What security methodologies and technologies can be applied to mitigating these threats?
- After the system has been implemented, how successful was threat modeling in mitigating attacks?

The last step is important because it provides input into modifying the organizational processes and choices of the underlying information system technology services (software, hardware, etc.) for the future development of the smart grid system.

Following an explicit methodology is crucial when conducting a threat modeling exercise. A number of threat modeling methodologies have been developed, all with their own strong and weak points. More details on specific threat modeling methodologies can be found in the references. The most important point, however, is that threat modeling is not simply an exercise in thinking about the problem but that the team involved in actually implementing and deploying the system is engaged and that the results are documented as part of the overall system architecture.

Threats are sometimes confused with risks, and some authors use the two terms interchangeably. A risk is the probability of a certain negative consequence given a particular action taken (or not taken) to mitigate a threat. Understanding the relationship between threats, possible mitigation actions, and the risks these actions entail is an important part of threat modeling, but the decision about what level of risk to accept and, therefore, what mitigation actions to take is often one for decision-makers and not system architects.

4.1.1 Trust Boundaries and Attack Surfaces

While different smart grid systems will differ in the details of their services, a couple of common threads run through all threat modeling exercises. Trust boundaries and attack surfaces are found in every information system architecture, and identifying them is the first step in determining where threats occur and, therefore, where security technologies need to be deployed.

A trust boundary is an interface where two principals with potentially different interests interact. An example is a customer accessing their billing information on the utility's website. The principals here are the customer and the utility's customer information system. The principals do not need to be people. Both principals could be smart grid services performing different functions, for example, a utility's advanced distribution management system (ADMS) accessing a customer's battery management system to schedule discharging for a peak event. The principals here are the ADMS system and the battery management system. Wherever there is a trust boundary, the principals on both sides of the trust boundary must prove their identity using a suitable authentication technique. In addition, the principal requesting access must be authorized to access specific information or services by the principal granting access.

An attack surface is the overall collection of interfaces where an attacker can compromise a system. Examples are:

- Weak or missing authentication leading to the compromised identity,
- Weak or missing authorization leading to compromised access,
- Lack of confidentiality protection on critical communications leading to compromised customer or company privacy,
- Key systems left unprotected from volumetric traffic increases leading to service interruptions due to denial of service attacks.

Trust boundaries and attack surfaces are related in that a trust boundary is often an attack surface. An unprotected trust boundary offers an attack vector for an attacker to exploit and enter the system or obtain data they are not privileged to obtain.

4.1.2   Basic Security Architectural Principles

There are a couple of basic architectural principles related to trust boundaries and attack surfaces. The first is to minimize the attack surface by reducing the number of external trust boundaries. This leads to fewer places where attackers can enter and compromise the system. Minimization should not, however, be done by arbitrarily consolidating system services behind an internal trust boundary. While conflicting interests are often clear between external principals and the smart grid system, such as a customer accessing a utility's website, they can often seem less clear internally. Internal principals such as employees of the smart grid operator working in different departments with responsibility for different services are all within the same organization. The tendency is to say that there is no trust boundary for internal principals. But the absence of a well-defined internal trust boundary could provide an opportunity for an attacker to broaden their attack if they do manage to break in.

This leads to the second principle, defense in depth. Defense in depth means to design the system so that security is not dependent on the integrity of a "hard" exterior shell (e.g., a firewall) while interaction between internal services is "soft," that is, left completely unprotected. An example of a soft interior is a service that accepts communication from other services without requiring authentication. If an attacker manages to penetrate the shell, then the system is entirely open for them to exploit. Proof of identity and the right to access should be required between internal principals and internal system services as well as between external principals and internal system services. Traffic between internal system services should also incorporate confidentiality protection. That way, if an attacker does manage to break in, their job becomes harder since they can't automatically leverage their original exploit to gain access to other parts of the system without more work. This gives the incident response team more time to detect and mitigate the attack.

4.2   Security Technologies

The Internet and Web communities have developed technologies to help in the design of secure systems dependent on the Internet for communication. Some of the technologies can also be used for securing documents and databases. This section contains an overview of these technologies. The details of the algorithms and protocols can be found in the references. At the end of the section, we briefly discuss the potential impact of quantum computing on security technologies.

4.2.1   Cryptographic Primitives

Security of communications over the Internet is built on top of the following cryptographic primitives [2]:

- One-way hash or digest functions that take as input a digital document or message and produce a highly compressed output,
- Symmetric, shared, or secret key cryptography in which two parties that want to confidentially communicate share a key for cryptographic operations,
- Asymmetric or public key cryptography in which the encrypted text, or ciphertext, is computed with a different key than is used to decrypt.

The algorithms behind these primitives utilize mathematical theorems from number theory and other areas to create a collection of standardized building blocks on which secure communication protocols and other operations are constructed.

Hash functions generate a fingerprint or digest of a digital document or message. Given a hash function, a document, and a digest claiming to have been created from the document, a receiver can verify the data integrity claim by applying the hash function to the document and comparing the received digest to the calculated digest. If they match, then the claim is correct. An important property of hash functions is that recreating the original text from the digest is impossible; in other words, the functions produce results in one direction only. In addition to verifying the data integrity of documents and messages, one-way hash functions are also used to keep passwords confidential while still allow easy comparison of the digests.

Some hash functions from the early years of the Internet, such as SHA-1 and MD5, have been found through cryptanalysis to have vulnerabilities and are not recommended today. SHA-256 and SHA-3, which are the subject of a National Institute of Standards and Technology (NIST) standards publications [3], were developed as replacements. Some smart grid security functions require specialized hash functions, for example, hashing passwords to keep them safe when they are stored in an identity management system database. The bcrypt algorithm [4], which was designed for keeping passwords confidential, has been subject to the most thorough cryptanalysis and is, therefore, a good choice for hashing passwords.

Symmetric or secret key encryption algorithms provide confidentiality protection when two principals have arranged to share a key. A key is a random-looking (to an eavesdropper) collection of bits that have been carefully selected according to the symmetric key algorithm in use to allow encryption and decryption. Encryption renders the digital document or message into a seemingly random stream of bits, while decryption converts the random bits into the original cleartext. This protects the text from an eavesdropper but allows the intended recipient, which possesses the key, to read it. Because the key both encrypts and decrypts, it must be kept secret.

There are two types of shared key encryption, stream ciphers, and block ciphers. Stream ciphers encrypt the text byte by byte, whereas block ciphers encrypt in large multi-byte chunks. As a practical matter, stream ciphers are not used in Internet communications except in a few minor niche applications. Block ciphers are the primary technology used in Internet communication and in smart grid security.

As with hash functions, early shared key encryption algorithms such as DES and Triple DES have been deprecated due to the discovery of vulnerabilities and increasing computing power allowing attackers to use unlimited computation to break the encryption (e.g., a "brute force" attack). In 2001, NIST sponsored a competition that ended in the selection of the Advanced Encryption Standard (AES) [5] for the new symmetric key cryptosystem standard. Since then, AES has been widely adopted worldwide. AES can be deployed with keys having sizes 128, 192, 256 bits; a 256-bit key provides maximum security and is recommended by the US government for encrypting documents having Top Secret status.

In contrast, asymmetric or public key cryptography uses two keys: one that the principal makes public and another that is kept secret. One key is used to encrypt while another is used to decrypt, and which key to use depends on the intended function. In comparison with shared key cryptography, public key cryptography is more computationally intensive. Many applications use public key cryptography to bootstrap a shared key, which they then use for securing further communication.

Public-key algorithms are based on difficult mathematical problems, for which, in most cases, no solution is currently known. There are a variety of public key algorithms, but the most commonly used are RSA [6], which stands for Rivest–Shamir–Adleman after the original authors, and ECC [6], or elliptic curve cryptography. RSA is based on the difficulty of finding prime factors given a large integer, while ECC is based on the algebraic structure of elliptic curves. Both RSA and ECC are in wide use today though the key size in RSA must be very large, 2048 bits minimum, to provide sufficient security, while a 256-bit key provides the equivalent level of security for ECC.

Key generation and length are important determinants of the strength of a cryptographic algorithm's security, both for symmetric and asymmetric cryptography. The length of a key should not be below the minimum value recommended for the algorithm, and in general, the longer the key, the stronger the security. During key generation, a pseudorandom number is often used to initialize the algorithm. The pseudorandom number generator should have sufficient randomness, or entropy, to ensure that weak keys are not generated. Many operating systems and language platforms neglect to incorporate a robust pseudorandom number generator.

In addition, some cryptosystems have special "weak" keys that decrease the computational effort required to break the encryption. AES, RSA, and ECC have the following properties with respect to weak keys:

- For AES, as long as the random number generator has sufficient entropy, there are no known weak keys.
- For RSA, as long as the key is at or above the minimum recommended key size of 2048 bits, the strength is sufficient.
- For ECC, as for RSA, a key size of 256 bits or above is sufficient. Some ECC algorithms are based on weak elliptic curves and should be avoided, but these cryptosystems have been weeded out from the NIST standards. As long as the cryptosystem in use follows NIST standards, it should be sufficiently strong.

However, in a legacy product, the cryptosystem may be based on a deprecated algorithm such as DES. A thorough understanding of which keys are weak is then required in order to avoid configuring the smart grid system with a weak key.

4.2.2   Cryptographic Operations

Cryptographic operations are built on top of the cryptographic primitives. A cryptographic operation is a procedure that has some security intent, for example, allowing the source of a message to be definitively authenticated and the integrity of the data during transit verified. In turn, cryptographic operations become part of the security protocols and security infrastructure that are the foundation of secure smart grid systems.

A particular problem unique to shared key systems is how both sides are provisioned with the same key. Public key systems don't have this problem since one key is designed to be made public. Any key-provisioning algorithm must ensure that the communication of a shared key is kept strictly confidential. In practice, there are many ways to do key provisioning. Here are a few commonly used techniques:

- The shared key is written on paper and must be typed in by the recipient or is incorporated into a physical piece of hardware, which is shipped to the installation to be physically installed in the device. The SIM card in cell phones uses this technique.
- A public key cryptosystem is used to bootstrap a shared key. If the public key algorithm in use is RSA, the generator of the shared key encrypts it with the recipient's public key, sends the encrypted key to the recipient, and the recipient decrypts with their private key.
- A key exchange algorithm, such as Diffie–Hellman key exchange [7], named after the authors of the algorithm, is used to generate a shared key. The Diffie–

Hellman algorithm is based on the same number theory approach as public key algorithms but on a different hard problem: the Diffie–Hellman problem. The Internet Key Exchange (IKE) protocol [8, 9], part of some smart grid standards, uses Diffie–Hellman for exchanging shared keys.

Another important cryptographic operation is the calculation of an authenticator on a document or message. Unlike a simple hash, an authenticator combines a hash calculation for data integrity verification with cryptographic operations for data origin authentication and non-reputation. If the cryptosystem uses asymmetric cryptography, the authenticator is called a digital signature. The sender calculates a digital signature by taking a hash of the digital material to be signed, then encrypting the hash with the its private key. The recipient hashes the received digital material using the same hashing algorithm, decrypts the signature sent by the sender using the sender's public key, and compares the calculated hash with the decrypted hash. If they match, then the authenticity of the message or document is verified.

While a shared key system such as AES can be used to calculate a message authenticator, the result does not have the non-reputation property of digital signatures. For digital signatures, the source of the message can be definitively identified if the sender includes their public key certificate in the message. Public key certificates are discussed in Section 4.3.1. For shared key systems, any party holding the shared key may have sent the message. Nevertheless, shared key authenticators, called message authentication codes (MACs), are useful in certain circumstances where non-repudiation is not important, and they are more computationally efficient than digital signatures.

A commonly used MAC calculation algorithm is hashed MAC or HMAC. HMAC [10] is a keyed cryptographic hash requiring both parties to share a key. As its name suggests, HMAC is paired with a hashing algorithm for computing the digest, in which case the combined algorithm is called HMAC-Z, where Z is the name of the hashing algorithm. An important property of HMAC is that it is immune to length extension attacks, as long as the hash algorithm is. In a length extension attack, the attacker takes the original message and the hash, extends the message and calculates a valid new hash based on the old one without knowing the key. The recipient then accepts the message as authentic. A MAC calculated with HMAC-SHA256 is not immune to length extension attacks, whereas HMAC paired with the newer SHA-3 standard [11] is. Consequently, MACs should be calculated with HMAC-SHA3.

Finally, encryption is important for maintaining the confidentiality of digital material, both when in motion, during communications, and when at rest, in files and databases on storage devices. While it is tempting to assume that a network run by the smart grid operator is a private network and therefore encryption on communication between devices within the network isn't necessary, this assumption can simplify an attacker's path through the network, should the external security perimeter be breached. If all internal communications are unencrypted, an attacker need only gain access to a single part of the network to eavesdrop on traffic between devices and between people and control systems. In addition, the privacy of customer data is extremely important, both from a legal and ethical standpoint. Keeping customer data encrypted in databases and files, therefore, is the best security practice. In addition, selected data about the smart grid system should be encrypted if an attacker could use the data to further an attack.

Either a shared key cryptosystem like AES or a public key cryptosystem like RSA with keys of suitable strength can serve as the basis of a solid encryption operation. Public key systems are somewhat easier to manage because key provisioning isn't required, but shared key systems are computationally more efficient. Most systems solve this tradeoff by authenticating the principals with digital signatures calculated using a public key cryptosystem and then bootstrapping a shared key using a key exchange protocol like IKE for encryption. If, however, a public key cryptosystem is used for encryption, the encryption/decryption process depends on the public key cryptosystem. For RSA, the use of the two keys is reversed from calculating a digital signature. The sending principal encrypts the digital material with the public key, and the recipient decrypts with its private key, rather than the other way around. For ECC, however, it is not possible to implement encryption/decryption directly. Instead, an algorithm similar to Diffie–Hellman, called the elliptic curve Diffie–Hellman (ECDH) [12, 13], is used to generate a shared key, and a shared key algorithm like AES is used to encrypt.

4.2.3 Security Protocols

The Internet protocol stack contains roughly four layers of interest to smart grid services: the link layer, the network layer, the transport layer, and the application layer. The lowest layer of the stack is the link layer. The link layer runs directly on top of the physical medium, like optical cables or wireless, which is why it is sometimes called the medium access (MAC) layer. The most common link layer protocol for smart grid applications is Ethernet, IEEE 802.1 [14] for wired links, or 802.11 [15] for wireless links, see Chapter 2 for more on smart grid link layer protocols. Addresses are only valid on the link between two routers or between a router and an end device, and packets (called frames at the link layer) need to be rewritten, have new source and destination addresses attached, and retransmitted when they traverse a router.

On the network layer, the Internet Protocol (IP), either IP Version 4 (IPv4) [16] or IP Version 6 (IPv6) [17], is the only network layer protocol in widespread deployment today. The network layer controls packet delivery between end devices, so in theory, IP addresses are valid end-to-end. In practice, if the smart grid deployment network runs IPv4 and the IP addresses in the organization's network are from the standardized private address space [18] (i.e., addresses having the form 192.168.x.y, 10.x.y.z, or 172.16.x.y-172.31.x.y) network address translation (NAT) will take place at the boundary of the smart grid network if the traffic is sent to or received from an external destination. NAT replaces a private address with the public address of the network address translation device or service and is used to hide the exact network topology from outside. IPv6 also has a local addressing standard, but application deployments on IPv6 are not as widespread as on IPv4.

Multiple protocols are in use at the transport layer depending on the application, but there are really only two of interest to smart grid applications: the User Datagram Protocol (UDP) [19] and the Transport Control Protocol (TCP) [20]. UDP is a simple connectionless protocol with minimal mechanisms, just checksums for data integrity and port numbers for addressing different applications at the same destination address. UDP provides no duplicate prevention, the guarantee of in-order delivery, or, for that matter, guarantee that the packets will be delivered at all. In contrast, TCP is a connection-based protocol in which the state of the connection between the two communication parties is maintained. TCP provides a reliable, in-order delivery with error checking end-to-end and includes a congestion avoidance protocol allowing applications to detect congestion and reduce their transmission rate if congestion is experienced on an intermediate link between the source and destination. Like UDP, TCP also supports port numbers for application delivery.

At the top layer are the application layer protocols. There are many application protocols, only a few of which are relevant for smart grid security. Those will be discussed in Section 4.3. One application layer protocol that is common to a broad variety of applications is the HyperText Transfer Protocol (HTTP) [21]. Client/server and web applications use HTTP to transfer client requests, typically from a browser, to the server and server responses to the client. But HTTP has outgrown its original purpose, which was simply to transfer hyperlinked documents from the server to the client, and today HTTP is used for a broad variety of purposes, including as a way to publish application program interfaces (APIs) to outside clients and partners. HTTP has evolved into a kind of generic application layer protocol.

While link layer security is important, especially for wireless links, the exact nature of the link layer security depends on the link layer standard, and the standards documents for the relevant link layer protocol in question should be consulted for details. On the other hand, IP and TCP/UDP are common to almost all applications and are more relevant to a smart grid security architecture. The following sections, therefore, describe the architecture of the security mechanisms for these layers, IP Security (IPsec) for IP and Transport Layer Security (TLS) for TCP/UDP. While security for HTTP is mostly handled by TLS via HTTPs, there is the provision in the HTTP standard for application-level security, allowing the sending party to include proof of authorization. HTTP application-level security is discussed in Section 4.3.2, where access control and authorization are presented.

4.2.3.1   IP Security (IPsec)

IPsec [22] supports two subprotocols for IP:

- The Authentication Header (AH) [23] protocol provides data origin authentication on an IP packet, including the IP header, which contains the source and destination address.

- The Encapsulating Security Payload (ESP) [24] protocol provides data origin authentication and both data origin authentication and confidentiality protection on the contents of the packet, but not the header.

In addition, if dynamic key provisioning is used, both protocols can also prevent retransmission of previously transmitted packets by an attacker, known as a replay attack. Packets protected with IPsec include a special IPsec header with control parameters.

There are two different deployment modes for the AH and ESP:

- Transport mode protects packets end-to-end. In transport mode, the IPsec header is inserted between the IP header and the transport layer header, which is usually TCP or UDP. Remote access virtual private networks (VPNs) are often deployed with transport mode IPsec.
- Tunnel mode protects the entire packet between one tunnel mode gateway and another. The tunnel packet consists of the following components in order: an outer IP header with the IP addresses of the sending and receiving gateways, the IPsec header, an inner IP header with the IP addresses of the sending and receiving end nodes, and the packet contents or payload. Tunnel mode IPsec is frequently used for site-to-site VPNs.

Two nodes that communicate with IPsec share a security association, which is a collection of session state applying to the unidirectional packet flow between one node and the other. Since the communication is usually bidirectional, two security associations are required. The security association consists of the following classes of state shared between the two endpoints:

- The subprotocol (either AH or ESP) and which deployment mode (either transport or tunnel) is in use,
- The cryptosystem (e.g., AES) used for the subprotocol,
- The keys for the cryptosystem.

The first step in protecting communications with IPSec is for the two sides to engage in an IKE transaction to establish a shared key and negotiate which cryptosystem will be used. Then the shared key, cryptosystem, and subprotocol are gathered into the security association. After the preliminaries are complete, the two sides can begin communicating with their chosen subprotocol.

4.2.3.2 Transport Layer Security (TLS)

TLS [25] provides security for transport layer protocols running on top of IP, most notably UDP and TCP. The TLS protocol consists of two subprotocols:

- The TLS Handshake protocol, a control protocol in which the security association for the session is established,
- The TLS Record protocol specifies the format of the protected data packets, and contains the transmitted data.

The security functions supported by TLS are similar to those supported by IPsec:

- Communication between the two parties is confidential. Confidentiality is maintained through encrypting the payload packets using a shared key cryptosystem. The shared key is negotiated during the establishment of the security association prior to sending any data.
- The parties to the communication are authenticated. In most applications of TLS, the server side authenticates itself to the client by providing a public key certificate, but the client is not required to authenticate to the server. However, in mutual TLS (mTLS), both the client and server exchange certificates and authenticate.
- Data integrity is ensured by calculating a MAC on the message using a MAC generation algorithm and shared key negotiated when the security association is set up.

The asymmetric nature of authentication in TLS stems from the original deployment intent: to secure sensitive transactions on the web. Customers of banks, e-commerce websites, and utilities needed a way to ensure that when they browsed a link to the corresponding website, they were not being spoofed. TLS accomplishes that by requiring the server to present an X.509 public key certificate created by a trusted certificate authority (CA). The browser contains a collection of such trusted CA certificates, which are used to validate the website's certificate. Section 4.3.1 discusses public key certificates and certificate authorities in more detail as they are key architectural components in managing identity and trust for services.

4.2.3.3 Remote Authentication Dial-in User Service (RADIUS)

While TLS requires public key certificates for authentication, certificates are optional for IKE/IPsec. If certificates are not used for IKE/IPsec, another option is to use the RADIUS [26] protocol. RADIUS transmits user and/or device identity, authorization, and key configuration between an identity and access management server, typically called an AAA (authentication, authorization, and accounting) server in the RADIUS standards, and the endpoint. The RADIUS protocol was originally designed to run over UDP transport and required the two endpoints to establish a security association based on pre-shared keys, which were only used for calculating a MAC. The specification made no provision for encrypting traffic between the client and server. Recently RADIUS was updated to allow TCP transport and TLS so the two sides can now set up a security association dynamically using certificates, and the traffic can be encrypted. If TLS is not used, RADIUS traffic between an endpoint and a RADIUS server should be encrypted using IPsec for confidentiality.

RADIUS messages are structured as a collection of key-value pairs. This provides RADIUS with maximum extensibility. RADIUS was originally designed many years ago for modem login to a single host computer over an analog telephone network, but it has been extended over the years. In addition to the key/value pairs specified in the original standards document, there are many extensions to RADIUS for various purposes, specialized deployment scenarios, etc., which have extended RADIUS into applications far beyond the original modem dial-in. An important extension is the Extensible Authentication Protocol (EAP) [27], which allows multiple types of authentication methods other than the simple user name and (cleartext) password specified by the original RADIUS login authentication message. EAP can be used with the IKE protocol to set up the security associations for IPsec.

While the smart grid security standards specify RADIUS for communicating with an identity and access management system, an updated version of RADIUS, whimsically called Diameter [28], was developed in the early 2000s and is required in the cellular telephony standards for authentication and authorization between a wireless network operator's identity and access management system and a user device, via a cellular base station. Diameter has the same basic key-value pair structure as RADIUS, but the design fixed some technical problems with RADIUS. Diameter is not directly backward compatible, but the important application messages in RADIUS have been duplicated, and, like RADIUS, many extensions have been standardized.

4.2.4   Impact of Quantum Computing

While the technological challenges of building a practical quantum computer have yet to be mastered, cryptographers and mathematicians have been researching quantum algorithms that would allow widely deployed cryptosystems to be broken. A well-known quantum algorithm called Shor's algorithm allows integer factorization in finite (polynomial) time. Unfortunately, because the security of many public key cryptosystems is based on the difficulty of factoring integers, the conclusion is that a practical quantum computer would have a huge impact on public key cryptosystems. Shor's algorithm makes it possible to break Diffie–Hellman as well as RSA and the commonly deployed ECC algorithms for any key size.

There are some known cryptosystems that have been proven safe against Shor's algorithm. One is supersingular isogeny Diffie–Hellman (SIDH) [30], a quantum-safe ECDH cryptosystem with a minimum quantum-safe key size of 2640 bits. The advantage of SIDH is that it uses much the same software infrastructure as existing ECC algorithms that are vulnerable, requiring fewer changes in operating systems and applications. Symmetric key algorithms are more resistant to attacks because they don't depend on integer factorization for security, so Shor's algorithm doesn't apply. The best-known attack using a quantum computer involves brute force and an algorithm called Grover's algorithm. Because a quantum computer search can be massively parallel, a brute force attack can be much faster than on a conventional computer. For AES or the keyed hashing algorithms used in calculating MACs, a key of size n bits provides $n/2$ bits of security against a quantum brute force attack, so the security can be easily increased by increasing the key size. Post-quantum cryptography is an active field of research [30], and new approaches such as lattice cryptography are under intensive investigation. NIST has yet to come up with recommendations in the area.

While the impact of quantum computing on existing cryptosystems would be huge, the amount of attention and research being invested in finding and implementing quantum cryptosystems seems to mitigate any immediate impact on smart grid deployments. Encryption is primarily used for communication, and most communication is ephemeral. By the time a practical quantum computer is built, network protocols are likely to have quantum-safe cryptography available. Encryption of data at rest in files and databases is somewhat more problematic, depending on how long the data is meant to be kept confidential. If there is a need to keep data confidential for longer than 8–10 years without reencryption, then a symmetric key algorithm like AES with a sufficiently long key is advisable to avoid any compromise should a quantum computer become practical.

One application of quantum technology that does not involve a fully functional quantum computer and therefore is practical today is quantum key exchange [31]. Quantum key exchange uses quantum superposition or entanglement to ensure that nobody eavesdrops on the exchange. To compromise the exchange, an attacker would have to make a measurement, thus causing collapse of the quantum state, which would be detectable by the two parties. The primary drawback of quantum key exchange is that it requires a standard authenticated channel between the two parties, which, if the channel is digital, means that either a symmetric or asymmetric key cryptosystem must be used to authenticate in setting up the channel. If the channel is analog, however, a classical cryptosystem isn't needed. An example of an analog channel is a dedicated fiber optic cable. Several commercial and government quantum key exchange networks with very limited distribution capability are in operation, but the impact on smart grid security systems seems remote.

4.3 Security Architecture and Infrastructure

Security architecture and infrastructure builds on top of the cryptographic operations from Section 4.2.2 to address threats identified during threat modeling. The components of a security architecture can be classified into five high-level categories:

- Managing the identity of principals,
- Controlling access of an authenticated principal to smart grid devices and services,
- Maintaining confidentiality and privacy of data and communication,
- Managing distributed denial of service (DDoS) attacks to ensure smart grid service quality,
- Detecting anomalies and distinguishing attacks from anomalies using machine learning (ML).

The following subsections examine these five functional areas in more detail.

4.3.1 Identity Management

In order for a principal to negotiate a smart grid service trust boundary, the identity of the principal must be established. In turn, a principal who seeks access to a smart grid service must have some way to establish the identity of the service. The nature of the principals on either side of the trust boundary doesn't really play a role. The same basic considerations apply whether the principal is a customer accessing a utility's website or a smart grid service accessing another smart grid service within the smart grid network.

However, in actual systems, different architectural approaches are brought to the problem of identity management depending on whether a person is involved or not. For example, when a customer browses to the utility's billing website, the browser sets up a TLS session and verifies the utility's service by checking whether the service's X.509 public key certificate is valid and was issued in the utility's name, thereby authenticating the utility's billing service to the browser and by extension to the customer. Once the session is set up, the customer is presented with a login page, where they are required to type in security credentials for a previously established account and authenticate themselves to the utility's billing service.

This example illustrates the two architectural approaches in widespread deployment for authenticating principals in smart grid systems:

- Public key certificates and certificate authorities, known as the public key infrastructure (PKI), identify servers and establish trust in the service between a service and a client. In some cases, bidirectional certificate exchanges are used when two services or automated agents need to establish mutual trust. The recipient of the certificate checks that the certificate was issued in the service's name, that the signature of the CA on the certificate is valid, and that the certificate has not expired or been revoked.
- Identity and access management (IAM) systems identify human principals (customers, employees of the organization running the smart grid service, etc.) to smart grid services through security credentials stored in a preestablished account. In most cases, the human principals are required to type in an account name and password to prove their identity, though some systems require a token generated by a token generator or biometric data like a fingerprint instead of a password. An additional authentication step called two-factor authentication (2FA) may also be required. The human principal is sent an access code, for example, to their cell phone and must type it in as an additional proof of identity.

The subsections below discuss these two architectural approaches to identity management.

There is an exception in the case of IAM systems. Some deployment technologies use IAM systems rather than a certificate for automated agents or services to identify themselves to other automated agents and services. The accounts held by these agents and services are then called service accounts. On the other hand, certificates are almost never used to identify people on the Internet, though they may be used in a corporate intranet setting to authenticate for VPN access and for signing and encrypting email if the company has deployed its own CA. The only reason for this asymmetry is that requiring everyone using the Internet to sign up and download a certificate, to say nothing of ensuring that the certificate is periodically renewed before it expires, was too big a task to undertake on the scale of the entire Internet in the early days of Internet commercialization. As a result, commercial services on the Internet began deploying traditional user name/password style login for identity management, borrowed from computer operating systems architectures of the early 1980s, and now this architectural pattern has become established.

4.3.1.1 Public Key Certificates and the Public Key Infrastructure (PKI)

A public key certificate is a collection of identity information on a principal, including the principal's name, which is often a domain name or service name, and their public key. The certificate is signed by a trusted third party, called a certificate authority (CA), attesting to the identity of the principal. The certificate may optionally include other information, such as a list of access privileges for the holder. Certificates for identity management of services that use IPsec and TLS are constructed according to the X.509 standard [32], and are therefore are also called X.509 certificates. Browsers and other applications that use certificates to authenticate a service are deployed with a cache of certificates from well-known certificate authorities, and certificates can be added to the cache. A CA constitutes a trust anchor, allowing the receiver of a service's certificate to transitively extend trust to the service. But how can the receiver of the certificate know that it can trust the CA?

The public key infrastructure (PKI) [33] was established to solve exactly this trust problem. The certificate of a CA must, in turn, be issued by and signed by another CA, and so on, up to the root CA, which acts as the root of trust. The chain of certificates from the principal to the root CA establishes a chain of trust. Certificate authorities closer to the root are generally more broadly deployed, with the root authority issuing certificates to a large number of other certificate authorities. Companies that specialize in network security, network operators, and others all have formed root authorities. Authorities closer to the root are therefore more likely to be in the application's cache because they are more broadly deployed.

Identity authentication using PKI proceeds through the following steps after the service sends its certificate to the application:

- Check that the CA certificate is in the cache. If not, request the service to provide it and validate it when received. This continues until a certificate in the chain is sent which was issued by an authority for which the application has a certificate in the cache.
- Check that the signature on the service's public key is valid.
- Check that the certificate has not expired.
- Check whether the certificate has been revoked. This may require an online request sent to the CA that issued the certificate, or it may require checking a

certificate revocation list (CRL) in the browser's cache. A CRL is a list of revoked certificates periodically distributed by some certificate authorities. Certificates can be revoked for any number of reasons, for example, if the service's private key is compromised.

If any of these checks fail, the application should reject the certificate and refuse to authenticate the service's identity.

4.3.1.2   Identity and Access Management (IAM) Services

An identity and access management (IAM) [34] service provides support for authenticating a principal, typically a person but also an automated agent, requesting access to other smart grid services. An IAM service consists of three components:

- A protocol for establishing an account. The account may belong to a person, or it may be a service account and belong to an automated agent,
- A directory service or database where the account identity and security credentials for the account identity are stored,
- A protocol for the principal to verify their identity when they want access to smart grid services and to grant the principal an access credential for ongoing access over a period of time.

For services designed to be accessed by people, an account is established through a special web page. The person provides an identifier for the account name and a password. The account name is usually a user name or email address for a person or a service name for a system service. The password should be of sufficient length (more than eight characters) and be sufficiently random to deter guessing attacks. Passwords should always be hashed using a suitable password hashing algorithm before being stored in the account database record. Many IAM systems use the LDAP directory service as a database for storing the account information, especially for internal accounts of employees. Often if the account is for a person, after establishing the account, the person receives an email at their email address with a link that they click through to verify their identity and activate the account. The rationale is that the person's email account provides definitive proof of their identity. Another way to establish proof of identity is by texting a pin to the person's cell phone number, as with 2FA for login. Neither method is foolproof of course, email and cell phone accounts can become compromised too.

   The protocol for a principal to verify their identity involves providing the account name and password. If the principal is a person, the account name and password are typed into a login web page. The same cryptographic hash function that is applied when the password is stored is then applied to the typed in password. The two are compared, and if they match, the principal's identity is established, or, alternatively, if 2FA is used, a pin is texted to the principal's cell phone number or sent in an email, and if the principal types in the texted pin, identity is established. Other security credentials that might be used to authenticate a human principal such as biometrics or a token generated by a token card follow roughly the same procedure, with the biometric data from the scan compared to that in the account database or the typed in token compared to the token generated by the IAM system's token generator. After a principal is authenticated, it is typically issued some kind of access credential granting access for a limited period of time. The level of access depends on the principal's authorization, the topic of the next section.

4.3.2 Access Control

Access control at the network and transport levels is provided by firewalls [35] and NATs. A firewall is a collection of allow/deny rules that control what packets are allowed through into the network behind the firewall. The allow/deny rules are based on a 5 tuple of fields in the packet header: the transport protocol type (e.g., TCP or UDP), the source IP address and transport protocol port, and the destination IP address and transport protocol port. The firewall rules can be configured to deny traffic except from particular IP addresses. For example, the rules in a firewall at a substation can allow unsolicited inbound traffic from addresses in the distribution system operations control center network VPN but deny such traffic from all other IP addresses. Similarly, private IP addressing can set up a routing perimeter beyond which traffic from an internal network cannot be routed out unless an address translation to a public address is installed in the NAT.

Authorization involves checking the access credentials to ensure that the authenticated principal has the right to access the service. Once authenticated, a principal is typically given access to the smart grid system service they are attempting to access. They must present an access credential to prove their right to access. Particularly for internal principals such as employees of the organization running the smart grid, finer-grained access control must be enforced to allow access only to principals with access privileges. An employee in the billing department should not be given access to the SCADA system, for example. If such fine-grained access control is not enforced, the identity credentials of an employee in one department who has had their account compromised could be used to access sensitive infrastructure services. Consequently, an additional authorization step may be required prior to providing access to a system service.

4.3.2.1 Role-Based Access Control (RBAC)

The most common technology for controlling fine-grained access is role-based access control (RBAC) [36]. With RBAC, principals are assigned particular roles having a collection of predefined access privileges. If the principal is a person, the role is assigned based on their job responsibility, position in the organization, or status as a customer with a particular kind of customer account (residential, business, etc.). Their access privileges are defined only for the specific services they need to accomplish their tasks, and their access credential is checked prior to access. In the example of a customer accessing their account at a utility's billing website, the customer is assigned a role of Customer, and the access privileges associated with the role allow an authenticated customer to read the billing data associated with their account and pay their bill. However, an authenticated customer can't read the billing data from another customer, nor can they access the SCADA system for substation control. If the principal is a service account, the role contains permissions allowing the automated agent authenticated to the account to access specific other services. An example is the privileges assigned to a principal with the role of Distribution System Control, allowing the ADMS service to access the AMI service for purposes of reading smart meter data. An RBAC system requires storing a principal's role along with their identity in the IAM directory service or public key certificate and requires a protocol that allows the accessed service to check the role based on the principal's access credential.

Roles should be defined according to the principle of least privilege. The principle of least privilege [37] states that only the minimum level of access privileges necessary for the principal to accomplish their task should be included in a role. If a principal, usually a person, has multiple tasks that they need to accomplish, then they should have multiple identity accounts. They could then authenticate with a different account when they want to switch tasks. The system could also support a way to switch roles when a principal wants to perform actions for another role. An example is an employee who is also a network administrator. The employee's basic identity should allow access to email, internal websites, etc., but they should be required to perform an extra authentication step should they want to change the routing tables on a router.

Access to all services, but especially critical infrastructure services, should be logged, and logs should be collected into a secure, tamper-proof, immutable audit logging service. Log entries should record the date and time access was granted, how long the access grant lasted, the identity of the principal requesting access, the principal's role, and what services and objects were accessed. In addition, unsuccessful attempts to access system services should be logged. Logs should be available dynamically to the incident response team so they can investigate the cause of an incident while it is underway. The audit logging service should allow regulatory authorities access in case an audit is required.

4.3.2.2   Access Credentials

Access credentials are commonly either included in a public key certificate or encoded in a time-limited access token granted by the IAM system. Less commonly, the access credentials may be provided as a separate attribute certificate.

If two endpoints are communicating with IPsec or mutual TLS and are using certificates to authenticate, then the access credentials in the certificate must be checked before an accessing principal is granted access, and the level of access opened should match the privileges given in the certificate. In addition, the signature or message authentication code on each message in the subsequent traffic flow must be verified to ensure that the traffic coming from the principal's IP address does, in fact, originate from a properly authorized principal. If one of the endpoints has authenticated with an IAM system and the endpoints are using TLS to secure traffic, the access credential typically takes the form of an access token that the principal embeds in subsequent application level protocol traffic. The TLS authenticator proves the identity of the accessing principal, and the access token proves their authorization.

Token-based authorization is heavily deployed by services that use HTTPS APIs to expose their functionality. HTTPS has a specific request header, the Authorization header [38], that allows the sender to provide an authentication token to the service. Traffic that incorporates a token must be encrypted as most types of authorization tokens are not cryptographically protected. There are a broad variety of token types, and which to include in a system architecture depends on the constraints imposed by the IAM service. For example, if a commercial off-the-shelf IAM service is deployed, then the token type used by other smart grid services to authorize must match the token type issued by the IAM service. Common token types are API keys, which do not use the Authorization header, bearer tokens, or OAuth tokens (OAuth is a particular authorization protocol used by many commercial IAM systems).

4.3.3   Confidentiality and Privacy

After a principal has established the right to access a system service, communication between the system service and the principal, whether a person or another system service, should be kept confidential by encrypting the exchanged messages. The confidentiality of data related to the operation of the smart grid system must be maintained to protect eavesdropping attackers from learning anything that could be used to propagate a successful attack. Exchanges with customers and the smart grid system must be similarly handled to protect customer privacy. For example, an attacker eavesdropping on traffic between a customer's smart thermostat and a utility's demand response control system may be able to learn when the customer is away from home, thereby providing the attacker with information to break into the customer's home without being detected. Finally, for legal and ethical reasons, customer data must be protected from unauthorized access, including by internal employees of the organization providing the smart grid service. Any of the cryptosystems discussed in Section 4.2.1 can be used with the protocols discussed in Section 4.2.3 to encrypt traffic between two endpoints, deployed according to the protocol's standard specification.

While confidentiality of data in motion is important, the confidentiality of data at rest is even more critical. If an attacker manages to gain access to the smart grid system, unencrypted data in a database or file can be stolen and mined for further attack targets, and because the data is sitting in a file system, the attacker has much more time to work out how to steal the data. Customer records are particularly vulnerable, but databases containing information on critical smart grid infrastructure are as well. One approach to maintaining the confidentiality of data at rest is to write the data in cleartext but store it on an encrypted disk partition. This approach only provides protection against the physical removal of the disk. If an attacker manages to compromise the operating system of any server with access to the encrypted partition, they can copy out the files and trawl through the files at their leisure. A better design is to encrypt individual records using a database with RBAC and read/write encryption so that the individual records are stored in an encrypted format. RBAC ensures that only authorized principals are allowed to access the data, and read/write encryption on the records ensures that, if a server with access to the data is compromised, the attacker will get nothing of value by copying out the database. With this design, an attacker must compromise the database system itself, a smaller and hopefully more hardened target, to gain access to cleartext data.

4.3.4 *Mitigating Distributed Denial of Service (DDoS) Attacks*

Some types of attacks aren't caused by failing to put the proper security architecture in place or neglecting some important security configuration during implementation or deployment but derive from attackers exploiting underlying features of the Internet architecture or even human nature. DDoS attacks [39] fall into this category. The immediate cause of a DDoS attack is an exponential volumetric increase in junk traffic coming from the Internet to the smart grid's Internet-facing services. The traffic has no intention of using the service; it is simply there to disrupt access. This increase causes real traffic from people and agents trying to actually use the service to slow down, which may, in turn, cause TCP sessions to time out. The end result is that the service becomes unusable. While such a result may be annoying for a customer browsing a utility website to view their bill, it threatens health and safety when control traffic to SCADA systems is disrupted.

DDoS attacks are enabled by a feature of the Internet routing system. When traffic is being routed over the Internet backbone, routers make no distinction based on the value of the traffic. Services determine the value of the traffic, and services in the Internet architecture are defined at the edge of the network. A router in the middle knows nothing about the nature of the service and what kind of traffic it will accept. Using various means, an attacker can construct a network, called a botnet, of thousands of compromised servers, virtual servers in cloud data centers, and devices. Once the botnet is in place, the attacker can then unleash it by sending instructions to bombard a smart grid service with traffic. Absent any information about the nature of the traffic; intermediate routers will continue to route the increased traffic despite the fact that the service becomes unusable.

DDoS attacks have been around for a while, and an extension to the Internet routing standards has been developed to help mitigate them. If an end system detects that it is getting junk traffic from an address, it can inject a message using the inter-domain routing protocol Border Gateway Protocol (BGP) upstream toward the subscriber edge router in their Internet service provider's (ISP) network. The message says in effect that the traffic is unwanted; please discard it. The subscriber edge router will then send the traffic down a black hole.

The key point, however, is how to identify unwanted traffic. Thankfully, there are now DDoS service providers that have developed DDoS mitigation services based on machine intelligence algorithms for identifying unwanted traffic. These DDoS services require that any traffic from the Internet to the smart grid service first go through the DDoS service, for example, resolving the DNS name for the smart grid service in a URL to the IP address of the DDoS service first, before the DDoS service redirects the traffic to the smart grid service. This provides the DDoS service an opportunity to examine the traffic in real time, and if it identifies any attack characteristics, to send the black hole message upstream to the ISP's subscriber edge router. This technology and business ecosystem around DDoS mitigation has reduced the threat of DDoS attacks considerably, but the attackers have now moved on to another such attack, which is discussed in the next section.

4.4   The Importance of Processes and People

Sometime during the spring of 2015, IT staff and system administrators at multiple distribution system operators in Ukraine were targeted by a spear-phishing attack [40]. In a phishing attack, an attacker sends an email that looks legitimate, but which contains a malicious link to a website or a document with an embedded malicious macro [41]. In either case, when the recipient clicks on the link, malware is installed on their computer that steals their authentication credentials, allowing the attacker access to the IT systems. A spear-phishing attack involves sending a phishing email to specific individuals targeted due to their status in the organization under attack in order to get account credentials with enhanced access privileges.

Several targeted individuals at three distribution system operators clicked on the link, giving the attackers, who are unknown but presumed to be working for the Russian military because Russia and Ukraine were involved in a border dispute at the time, complete access to the affected companies' IT systems, but not yet to the SCADA systems. The attackers spent the next 6 months rooting around in the distribution operators' systems: downloading the credentials for workers with access to the SCADA VPNs from the Windows Active Directory databases, installing malicious firmware to disable the Ethernet to serial converters in substations, and reprogramming the backup power to two of the control centers so they could be disabled.

On December 23, 2015, the attackers struck. Using the VPN credentials, they disabled the UPS, opened all the breakers, and disabled the SCADA system through the malicious firmware. In addition, they wiped all the operator hard drives clean and launched a telephone denial of service attack on the customer support center with calls appearing to come from Moscow so nobody could report it. In total, 230,000 households lost power as a result. Fortunately, the distribution system operators involved had recent backups of the hard disks and extensive audit logs, allowing them to trace what modifications the attackers had made to the smart grid system and restore the hard drives. They were able to restore power within 6–8 h.

This incident, and many similar incidents since, points up a critical missing piece: the importance of cybersecurity best practices [42] by having processes in place and training people to respond to and deal with an attack. Involving the people who are managing the smart grid service in security and having processes in place to deal with incidents are critical to maintaining a safe and reliable smart grid service, and break down into the following three phases:

- Training people who manage the critical IT and SCADA systems and putting processes in place to engage in security best practices,
- Determining how to respond when an attacker has breached the defenses and is detected,
- Conducting a postmortem analysis after the attack has been resolved and the service restored.

With respect to the first point, phishing attacks are a kind of psychological warfare that many companies counter by running fake phishing campaigns. Every 6 months or so, a fake phishing email is sent to employees, and those that click on the link get a response email that tells them they've been spoofed. If an employee is taken in by the fake phishing attack multiple times, they are required to attend training in how to detect a phishing email and what to do if one shows up. Likewise, the fact that the attackers in the Ukrainian distribution operator attack were able to get user names and passwords from the Active Directory service suggests that, at minimum, the Active Directory records were not encrypted and that the passwords were not even hashed. As discussed in Section 4.3.1.2, storing hashed passwords is a security best practice for preventing attackers from harvesting more passwords than those from the accounts that they compromise.

A particularly difficult kind of attack to mitigate is a ransomware attack [43]. The attackers use a phishing attack to steal credentials from an employee or find some other way to break into the smart grid organization's IT system. They then encrypt all of the file systems in the organization, making them unreadable. Finally, they send an email to the organization management demanding a ransom, usually in bitcoin, for the key. In recent years, ransomware attacks have become major business operations, with companies formed to organize the attacks, like organized crime syndicates and victim companies paying ransoms upwards of $10 million. Regularly scheduled backups stored offline in an archive are another best practice that can help a smart grid organization recover from a ransomware attack. While restoring all systems from backups is time-consuming, paying the ransom will only encourage the attackers to continue, and there is no guarantee that the attackers will actually provide a functional key.

Cybersecurity best practices require that all employee computers, including laptops, have malware protection software installed, that someone in the security team tracks notifications on security patches and updates, and that they record when updates are actually installed to ensure that the installation actually happens. Security updates for SCADA, ADMS, and other software that directly manages the smart grid systems should also be tracked by the security team, though installation may require collaboration with the operations team who are more familiar with the respective management systems. Another good practice is to run a penetration test, or pen test, periodically and especially after installing updates of major software systems or making major configuration changes, particularly in systems on the security perimeter like firewalls and routers. Pen tests are typically run by an outside company to avoid any insider bias. A pen test runs through a variety of known configuration vulnerabilities to determine if the smart grid system leaves any openings for attack. If any vulnerability is found, the security team can fix it.

Handling an attack that is underway is typically the job of the incident response team [44]. The incident response team must determine whether the attacker is still active in the network and cut them off if so, then go back through the audit logs and determine what systems were compromised and restore them. A regularly scheduled, periodic review of audit logs, either by a person or by an ML program designed to spot anomalies, could help to spot an attack before it becomes serious. Such a review process may have caught the Ukrainian attack prior to the attackers triggering major damage since the attackers spent a considerable amount of time setting up the attack.

Finally, as discussed in Section 4.1.1, an objective and blame-free postmortem is extremely important in improving the smart grid system and preventing another attack. If defects in the security technology or deployment are found, for example, lack of a firewall at a trust boundary, then these can be remedied by either

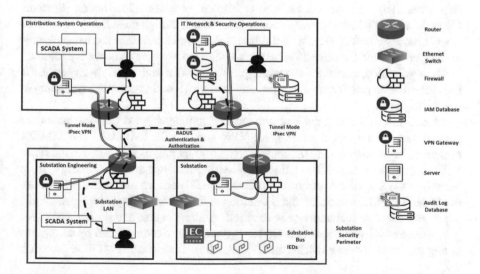

Fig. 4.1 Example security reference architecture for a substation control network

implementing the changes or bringing in a vendor to provide the necessary technology. If some problem occurred in the preattack or attack phase as part of the incident response, then the necessary changes can be made in processes or employee training to ensure that everyone is ready the next time around. A postmortem needs to be conducted in a way that encourages everyone to examine their role in the incident and to acknowledge what they can do better, without fear of retribution or blame.

4.5   Smart Grid Security Reference Architecture

A security reference architecture summarizes in one diagram the technical measures taken to address the identified threats [45]. Figure 4.1 shows an example, a security reference architecture for a smart grid deployment connecting a management system to a substation. Only one substation is shown. Distribution Systems Operations and IT Network and Security Operations are two networks outside the substation security perimeter that need access to the substation network. Within the substation security perimeter, Substation Engineering handles deployment and operations for the network of substations. The SCADA system is shown as split between Substation Engineering and Distribution System Operations. The substations themselves internally run a substation bus that conforms to the IEC 61850 communications standard. Various IEDs (transformers, circuit breakers, capacitors, etc.) are connected to the bus. The connection between Substation Engineering and the Substation is at the link layer, over the Substation LAN, so no router is required.

Because the intent of this diagram is to highlight the security architecture of the substation network, no connection is shown between Distribution Systems Operations and IT Network and Security Operations, though most likely one would exist. Likewise, no connection to the Internet is shown between the two management networks and the substation networks, though they may run over the Internet, which is why there are firewalls on the egress routers for both sides. In any case, the firewall rules are programmed to reject any traffic except that coming from/going to the VPN gateway.

In addition to the security and networking appliances, two protocol paths are shown: one for the tunnel mode IPsec VPN and one for RADIUS. The SCADA system and operators in the Distribution Systems Operations network use the IPsec VPN to connect into the Substation Engineering network and, if necessary, from there into the Substation network. Similarly, the IT network and security team use the IPsec VPN to connect into the Substation and Substation Engineering networks when any network maintenance is required. Authentication and authorization for grid operators, IT personnel, and service processes are accomplished by communicating the account credentials to the IAM database through the RADIUS protocol, and credentials providing RBAC authorization are returned if the account credentials are authenticated. A centralized audit logging database is also shown. Audit log files from individual devices are collected and archived periodically through the IPsec VPN. Centralized databases are specified in the architecture for the IAM system and audit logs to avoid a fragmented collection of files in different networks that would be difficult to defend against an attacker and could slow down an investigation by the incident response team during an incident.

4.6 Summary

In this chapter, we've taken a high-level look at security architecture for smart grid systems. Designing a security architecture starts with a threat model. From the threat model, trust boundaries can be derived, and an attack surface defined. Consolidating the external trust boundaries to minimize the attack surface can help reduce opportunities for attackers to enter, but such consolidation should not be at the expense of properly securing internal traffic. These two considerations lead to two basic security architectural principles: reduce the attack surface and defense in depth.

We then examined the cryptographic primitives available for building secure communications and data storage. These consist of hashing, symmetric-key cryptography, and asymmetric or public key cryptography, and we reviewed the advantages and disadvantages of the currently most widely deployed algorithms. Cryptographic primitives are used to build cryptographic operations, and we discussed the important ones: key provisioning and exchange, calculation of keyed message authentication codes, and encryption. We then examined IPsec, TLS, and RADIUS, which all play important roles in smart grid security standards. We also briefly looked at the impact of quantum computing on smart grid security. And finally, security appliances and systems are built on top of security protocols, and we discussed systems for identity management, access control, and controlling DDoS attacks.

But security systems are really only as good as the people and processes that maintain and use them. Using the 2015 attack on the Ukrainian distribution grid as a motivating example, we discussed some security best practices for training personnel to recognize phishing attacks and other procedures to protect against attacks and help deal with an attack and its aftermath, should one occur. The last section contains a simple sample security architecture for a substation network in a distribution system operator and shows how the different technologies work together to maintain overall security.

References

1. Shostack, A., *Threat Modeling: Designing for Security*, 624 pp. Wiley, New York, 2014.
2. Kempf, J., *Wireless Internet Security: Architecture and Protocols*, 212 pp. Cambridge University Press, Cambridge, UK, 2008.
3. National Institute of Standards and Technology, "Secure Hash Standards (SHS)," FIPS PUB 180-4, 31 pp. Information Technology Laboratory, Gaithersburg, MD, 2015.
4. Provos, N., and Mazières, D., "A Future-Adaptable Password Scheme". [Online]: https://www. usenix.org/legacy/events/usenix99/full_papers/provos/provos.pdf (Accessed 2020-08-21).
5. National Institute of Standards and Technology, "Specification for the Advanced Encryption Standard (AES)", FIPS PUB 197, 47 pp. Information Technology Laboratory, Gaithersburg, MD, 2001.
6. National Institute of Standards and Technology, "Digital Signature Standard (DSS)", FIPS PUB 186-4, 121 pp. Information Technology Laboratory, Gaithersburg, MD, 2013.
7. Rescorla, E., "Diffie-Hellman Key Agreement Method", RFC 2631, Internet Engineering Task Force, June, 1999.
8. Kaufman, C., "Internet Key Exchange (IKEv2) Protocol", RFC 4306, Internet Engineering Task Force, December, 2005.
9. Hoffman, P., "Algorithms for Internet Key Exchange version 1 (IKEv1)", RFC 4109, May 2005.
10. National Institute of Standards and Technology, "The Keyed-Hash Message Authentication Code (HMAC)", FIPS PUB 198-1, 7 pp. Information Technology Laboratory, Gaithersburg, MD, 2008.
11. National Institute of Standards and Technology, "SHA-3 Standard: Permutation-Based Hash and Extendable-Output Functions", FIPS PUB 202, 29 pp. Information Technology Laboratory, Gaithersburg, MD, 2015.
12. Haakegaard, R., and Lang, J., "The Elliptic Curve Diffie-Hellman (ECDH)". [Online]: http:// koclab.cs.ucsb.edu/teaching/ecc/project/2015Projects/Haakegaard+Lang.pdf (Accessed 2020-08-21).
13. Herzog, J., and Khazan, R., "Use of Static-Static Elliptic Curve Diffie-Hellman Key Agreement in Cryptographic Message Syntax", RFC 6278, Internet Engineering Task Force, June 2011.
14. IEEE, "IEEE 802.1". [Online]: https://1.ieee802.org/ (Accessed 2020-08-21).
15. IEEE, "IEEE 802.11". [Online]: https://www.ieee802.org/11/ (Accessed 2020-08-21).
16. Postel, J., et. al., "Internet Protocol", STD 5, Internet Engineering Task Force, September, 1981.
17. Deering, S., and Hinden, R., "Internet Protocol, Version 6 (IPv6) Specification", RFC 2460, Internet Engineering Task Force, December, 1998.
18. Rekhter, Y., et al., "Address Allocation for Private Internets", RFC 1918, February, 1998.
19. Postel, J., "User Datagram Protocol", STD 6, Internet Engineering Task Force, October, 1980.
20. Postel, J., "Transmission Control Protocol", STD 7, Internet Engineering Task Force, September, 1981.
21. Belshe, M., Peon, R., and Thomson, M., "Hypertext Transfer Protocol Version 2 (HTTP/2)", RFC 7540, Internet Engineering Task Force, May, 2015.
22. Kent, S., and Seo, K., "Security Architecture for the Internet Protocol", RFC 4301, Internet Engineering Task Force, December, 2005.
23. Kent, S., "IP Authentication Header", RFC 4302, Internet Engineering Task Force, December, 2005.
24. Kent, S., "IP Encapsulating Security Payload (ESP), RFC 4303, Internet Engineering Task Force, December, 2005.
25. Dierks, T., and Rescorla, E., "The Transport Layer Security (TLS) Protocol Version 1.1", RFC 4346, Internet Engineering Task Force, April, 2006.
26. Rigney, C., et. al., "Remote Authentication Dial In User Service (RADIUS)", RFC 2138, Internet Engineering Task Force, April, 1997.
27. Aboba, B., et. al. "Extensible Authentication Protocol (EAP)", RFC 3748, Internet Engineering Task Force, June, 2004.

28. Fajardo, V., et. al., "Diameter Base Protocol", RFC 6733, Internet Engineering Task Force, October, 2012.
29. Jao, D., and De Feo, L., "Towards quantum-resistant cryptosystems from supersingular elliptic curve isogenies". [Online]: https://web.archive.org/web/20120507222310/http://eprint.iacr.org/2011/506.pdf (Accessed 2020-08-21).
30. Wikipedia, "Post-quantum cryptography". [Online]: https://en.wikipedia.org/wiki/Post-quantum_cryptography (Accessed 2020-08-21).
31. Wikipedia, "Quantum key distribution". [Online]: https://en.wikipedia.org/wiki/Quantum_key_distribution (Accessed 2020-08-21).
32. Russell, A., "What is an X.509 Certificate?". [Online]: https://www.ssl.com/faqs/what-is-an-x-509-certificate/ (Accessed 2020-08-21).
33. Polk, T., "Introduction to Public Key Infrastructure". [Online]: https://ncvhs.hhs.gov/wp-content/uploads/2014/05/050113p3.pdf (Accessed 2020-08-21).
34. Wikipedia, "Identity management". [Online]: https://en.wikipedia.org/wiki/Identity_management (Accessed 2020-08-21).
35. Forcepoint, "What is a Firewall?". [Online]: https://www.forcepoint.com/cyber-edu/firewall (Accessed 2020-08-21).
36. Auth0, "What is Role Based Access Control?". [Online]: https://auth0.com/docs/authorization/rbac (Accessed 2020-08-21).
37. CSIA, "Least Privilege". [Online]: https://us-cert.cisa.gov/bsi/articles/knowledge/principles/least-privilege (Accessed 2020-08-21).
38. Mozilla Foundation, "Authorization – HTTP". [Online]: https://developer.mozilla.org/en-US/docs/Web/HTTP/Headers/Authorization (Accessed 2002-08-21).
39. Cloudflare, "What is a DDoS attack?". [Online]: https://www.cloudflare.com/learning/ddos/what-is-a-ddos-attack/ (Accessed 2020-08-21).
40. Greenberg, A., "How an Entire Nation Became Russia's Test Lab for Cyberwar". [Online]: https://www.wired.com/story/russian-hackers-attack-ukraine/ (Accessed 2020-08-22).
41. Fruhlinger, J., "What is phishing? How this cyber attack works and how to prevent it". [Online]: https://www.csoonline.com/article/2117843/what-is-phishing-how-this-cyber-attack-works-and-how-to-prevent-it.html (Accessed 2020-08-22).
42. Mullane, M., "The five pillars of cyber security". [Online]: https://medium.com/swlh/the-five-pillars-of-cyber-security-d247cd2e49cb (Accessed 2020-08-22).
43. Fruhling, J., "Ransomware explained: How it works and how to remove it". [Online]: https://www.csoonline.com/article/3236183/what-is-ransomware-how-it-works-and-how-to-remove-it.html (Accessed 2020-08-22).
44. Matthews, T., "Organization: How to Build an Incident Response Team". [Online]: https://www.exabeam.com/incident-response/csirt/ (Accessed 2020-08-22).
45. NIST, "Guidelines for Smart Grid Cybersecurity: Volume 1-Smart Grid Cybersecurity Strategy, Architecture, and High-Level Requirements", NISTIR 7628 Revision 1, September, 2014.

Chapter 5
Energy Internet of Things

Prior to and during the commercialization of the Internet in the early 1990s, most of the focus was on supporting end devices that were used by humans. While there were some early experiments on connecting machines to the Internet as end devices, the real growth in communication between two devices had to wait until device miniaturization, low-power electronics, and wireless link technology were far enough along that connectivity could be incorporated into practically any device that needed it. This trend is called the "Internet of Things" (IoT). IoT is a superset of Machine-to-Machine (M2M) communication, encompassing the interconnection of intelligent devices and management platforms through advanced communication technologies developed for the Internet.

A device that can be connected using IoT is often referred to as a "smart object," that is, an object that can be accessed from anywhere at any time. Smart objects can range from ordinary sensors to sophisticated industrial devices and include home appliances, light switches, load controllers, thermostats, electric grid devices, environmental sensors, wearables, cameras, cars, infrastructure in homes and cities, and more. These devices connect to the Internet, often through some intermediate network for access control, to transmit data about their state or the state of the environment in which they are embedded and to receive data controlling the device's state. IoT brings the Internet, data collection, processing, and analytics to the real world of physical objects.

This chapter focuses on the energy Internet of Things (IoT) with trending communication technologies in IoT networks, IoT devices, and sensors, and also discusses IoT applications in the smart grid environment, sometimes called the "Energy of Internet of Things." Section 5.1 describes the Internet of Things (IoT) and Industrial Internet of Things (IIoT), while Section 5.2 discusses communication and software concepts associated with IoT devices. In Section 5.3, the trending communication technologies in IoT networks are discussed. Section 5.5 contains an in-depth study of an Energy IoT use case, electric vehicle charging, showing how

U. Cali et al., *Digitalization of Power Markets and Systems Using Energy Informatics*, https://doi.org/10.1007/978-3-030-83301-5_5

Energy IoT can provide valuable services for consumers and utilities that would be otherwise difficult to implement. Finally, Section 5.6 presents the chapter summary.

5.1   What Are the Internet of Things (IoT) and the Industrial Internet of Things (IIoT)?

The idea of connecting machines has been around in the industry for longer than IoT under the term Machine-to-Machine (M2M) communication. M2M communication has been widely used to allow two or more machines to communicate without requiring any human interaction [1]. But industrial M2M devices were based on technology that was too expensive for incorporation into consumer devices. In consumer markets, IoT-based applications have become more common and practical through the general development of inexpensive Internet technologies such as access to low-cost, low-power sensors, connectivity, cloud computing platforms, mobility, machine learning, analytics, and conversational artificial intelligence.

Prior to and during the commercialization of the Internet in the early 1990s, most of the focus was on supporting end devices that were used by humans. While there were some early experiments on connecting machines to the Internet as end devices, the real growth in communication between two devices had to wait until device miniaturization, low-power electronics, and wireless link technology were far enough along that connectivity could be incorporated into practically any device that needed it. This trend is called the "Internet of Things" (IoT). IoT is a superset of Machine-to-Machine (M2M) communication, encompassing the interconnection of intelligent devices and management platforms through advanced communication technologies developed for the Internet.

A device that can be connected using IoT is often referred to as a "smart object," that is, an object that can be accessed from anywhere at any time. Smart objects can range from ordinary sensors to sophisticated industrial devices and include home appliances, light switches, load controllers, thermostats, electric grid devices, environmental sensors, wearables, cameras, cars, infrastructure in homes and cities, and more. These devices connect to the Internet, often through some intermediate network for access control, to transmit data about their state or the state of the environment in which they are embedded and to receive data controlling the device's state. IoT brings the Internet, data collection, processing, and analytics to the real world of physical objects.

This chapter focuses on the energy Internet of Things (IoT) with trending communication technologies in IoT networks, IoT devices, and sensors, and also discusses IoT applications in the smart grid environment, sometimes called the "Energy of Internet of Things." With these technologies, IoT devices can automatically collect and share data with minimal human intervention. IoT cloud-based platforms allow automatic recording and monitoring of data and the adjusting of interaction between connected things and their environment. Many sources expected that by 2025 over 22 billion IoT devices would be deployed worldwide [2].

The first IoT implementation was a vending machine at Carnegie Mellon that was connected to ARPANET in the 1980s. However, the term Internet of Things was invented by British technologist Kevin Ashton in 1999 [3]. With the advance of low-cost and low-power technology for implementing sensors and wireless communications, incorporating IoT functionality into devices for markets outside of industrial equipment became possible. IoT technology relies on IP-based networks to transfer data from the device to a cloud or middleware platform [4]. IoT devices can automatically collect and share data with minimal human intervention. IoT cloud-based platforms allow recording and monitoring the collected data and adjusting the interaction between connected devices and between the devices and their environment.

While consumer markets have benefited from low-cost IoT devices, the technology for low-cost IoT has found its way into industrial markets as well. Industrial IoT (IIoT) is a subset of but also the biggest and most important market for IoT. IIOT refers to IoT technology in industrial applications, including instrumentation and control of sensors and devices connected through the Internet protocol to industrial enterprise networks, though typically not connected directly to the public Internet. IIoT has a seen fast adoption with industry support and incentives as compared to consumer IoT applications. IIoT is an advanced version of Machine-to-Machine communication (M2M), sometimes called the fourth wave of the industrial revolution, or Industry 4.0. The goal of IIoT and Industry 4.0 is to achieve full remote monitoring and control of industrial processes through Internet protocol-based connectivity, cloud, and data analytics technologies. IIoT has been used in many industrial fields, such as smart manufacturing, preventive and predictive maintenance, smart cities, smart grids, connected and smart logistics, supply chain monitor, and management systems. IIoT applications provide many advantages to industry, primarily reduced costs, increased productivity, and efficiency [5]. According to the research released by IDC, the industrial IoT market is estimated to reach $123.89 billion by 2021. Just as with enterprise IT applications, industrial IoT provides industrial businesses with important leverage to help them make business decisions faster and more accurately and allows them to understand their business processes better [6].

Figure 5.1 shows the IoT concept for consumer and industrial applications.

5.2 Communication and Software Concepts for IoT Devices

IoT sensors play an important role in creating solutions using IoT. IoT sensors detect changes in the external environment and relay data on the change to data storage and analytics applications, typically in servers and cloud platforms at remote sites. There are a variety of IoT sensors and devices used to detect and measure various physical properties, such as temperature, humidity, acceleration, gyro, pressure, speed, energy, CO_2, and many more. Software platforms connect the edge devices

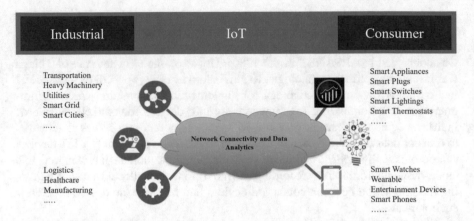

Fig. 5.1 IoT concept for consumer and industrial applications

and communication networks to the end-user applications for data analytics, display, and storage. IoT software addresses the key areas of networking and action through platforms, embedded systems, partner systems, and middleware. These individual and master applications are responsible for data collection, device integration, and data analytics [7].

The data collection process is conducted by managing measurements, data filtering, data security, and aggregation of data through data exchange protocols. The measured and formatted data is collected from multiple devices and distributed in accordance with the configuration at the device and application platform. Data integration or synthesis combines data from different IoT devices to provide users with a comprehensive visualization of the environment or process being monitored. Software applications on the IoT platform can analyze and visualize, allowing the user to understand historical data and predict future outcomes. In addition, real-time analytics can utilize IoT-based solutions to track the performance of machinery, provide predictive maintenance recommendations, and better understand data related to their devices, such as temperature, humidity, light levels, energy consumption or generation, and CO_2 concentrations or production.

Connectivity is a crucial factor required for the success of IoT applications and requires high reliability and availability. Many IoT applications, particularly for the utility and other markets where the devices are spread across large areas, focus on cellular technologies, such as 4G LTE-M and NB-IoT cellular wireless connectivity technologies, if the applications require widespread availability of 4G through public access wireless carriers. IoT applications with a requirement for wide area coverage can utilize low-power WAN (LPWNA) technologies, such as LoRaWAN or SigFox. Applications for consumer or enterprise markets where short-range coverage is required can utilize Wi-Fi, ZigBee, Z-Wave, Thread, Bluetooth Low Energy (BLE), 6LoWPA, EnOcean, and others [8]. Figure 5.2 illustrates a high-level architecture for platform systems supporting IoT devices.

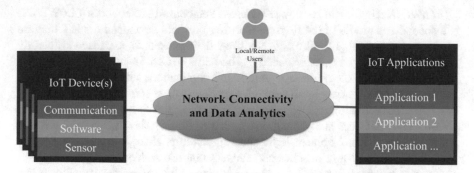

Fig. 5.2   The architecture of communication and software concepts associated with IoT devices

5.3   Communication Technologies for IoT Applications

In this section, communication technologies for IoT-enabled applications are summarized. These technologies include Wi-Fi Direct, Bluetooth Low Energy (BLE), ZigBee, Z-Wave, 6LoWPAN, EnOcean, Thread, ANT+, Near-Field Communication (NFC), LoRa, SigFox, Narrowband-IoT (NB-IoT), and LTE-M.

Wi-Fi Direct: Wi-Fi Direct is a communication technology developed and incorporated in devices in the early 2000s, which allows for device-to-device communication and eliminates the need for an access point. It can support the data rate of Wi-Fi, but with lower latency. In a Wi-Fi Direct network, a device acts as an access point, while the other device connects to it using the Wi-Fi Protected Setup (WPS) and Wi-Fi Protected Access (WPA/WPA2) security protocols [9].

Bluetooth Low Energy (BLE): Bluetooth Low Energy (BLE) is a short-range wireless communication technology that runs on low power and is known as Bluetooth 4.0. BLE technology is widely used for applications needing low amounts of data and can therefore run on battery power for years without having to change the battery. BLE data transfer is one-way only [10]. There are many everyday devices, such as smartphones, smartwatches, fitness trackers, wireless headphones, and computers that use BLE technology.

ZigBee: ZigBee is a popular low-power wireless local area network technology, which was specifically designed for networks needing to transfer a small amount of data using very little power, with most connected devices running off a battery. ZigBee also supports a mesh network specification where devices are connected with many interconnections between other network nodes [11]. ZigBee is a good candidate for IoT and IIoT applications because requiring low-power and mesh network support.

Z-Wave: Z-Wave is a short-range wireless communication technology that was developed for smart home networks. A Z-Wave network consists of IoT-enabled devices with a controller that acts as a gateway to a network running the Internet protocol connecting to the public Internet. Devices can connect to the Internet through the controller. A Z-Wave device can last up to 10 years on a coin cell battery due to its low power consumption.

6LoWPAN: IPv6 over Low power Wireless Personal Area Networks (6LoWPAN) is a specialized profile of the IPv6 networking protocol developed for IoT that are resource-constrained [12]. 6LoWPAN allows IPv6 packets to be carried efficiently within small link-layer frames defined by the IEEE 802.15.4 standard. Its characteristics make the technology ideal for generic IoT applications with Internet-connected devices. It supports a large mesh network topology, robust communication, and very low power consumption.

EnOcean: EnOcean is a popular energy harvesting wireless technology that can generate energy from ambient light and temperature. Energy harvesting enables wireless and battery-less switches and sensors that use ambient energy as a power source for battery-less wireless communication. The battery-less operation can reduce installation and maintenance costs. With energy harvesting technology, sensors can communicate powered by EnOcean technology without any concern about changing batteries.

Thread: Thread is a short-range wireless communication technology based on IEEE 802.15.4 and 6LoWPANand was designed for connecting products around the home and in buildings to each other, the Internet, and the cloud. It can enable secure, lower power reliable communications for low-power IoT applications [13]. A Thread network is simple to install, highly secure, scalable to hundreds of devices, and developed to run on low-power IEEE 802.15.4 chipsets. It is independent of other 802.15 mesh networking protocols, such a ZigBee, Z-Wave, and BLE.

ANT+: ANT is an open-access multicast wireless sensor network (WSN) protocol that supports ultra-low power IoT applications. ANT+ is the successor to the basic ANT protocol with additional interoperability functionalities, such as the collection, tracking of sensor data, and automatic transfer. ANT+ supports various network topologies such as peer to peer, tree, star, and mesh topologies, and network to network connections. ANT and ANT+ networks are used in IoT applications [14].

Near-Field Communication (NFC): Near-Field Communication (NFC) is a short-range wireless technology in the 13.56 MHz band for Radio-Frequency Identification (RFID). NFC is used for the exchange of information between two devices. RFID supports a bidirectional radio frequency identification system consisting of tags and readers that can be interfaced with handheld computing devices or personal computers. NFC consists of communication protocols for electronic devices, typically a mobile device and a standard device. With advantages such as simplicity, security, and connecting unconnected devices, NFC has the potential to significantly improve IoT adoption since it simplifies connecting two different IoT devices [15].

LoRa: LoRa is a low-power and wide area wireless technology (LPWAN). LoRaWAN developed based on LoRa wireless and is a well-known open standard protocol designed to enable communications between battery-powered IoT devices that require a low data rate and the Internet. Generally, LoRaWANs are connected using a star topology, with the central node serving as a gateway. LoRa supports long-range, low-cost, mobile, and secure end-to-end bidirectional communication for IoT applications. LoRaWAN technology is specifically designed for IoT applications and allows cost-efficient deployments and operations for both public and private networks [16].

SigFox: SigFox is a long-range, low power, low data rate LPWAN technology that was developed to provide wireless connectivity for devices like remote sensors, actuators, and IoT devices. SigFox acts as a cellular network operator and provides a public network solution for low-throughput IoT applications. The SigFox protocol has been developed to limit the amount of power required by using an ultra-narrow frequency band. Remote IoT devices can operate on battery power for very extended periods without the need for any battery changes or maintenance [17].

LTE-M: Another LTE-based technology developed for IoT use is LTE-M. LTE-M was developed before NB-IoT, as part of a suite of protocols introduced by 3GPP [18], including LTE category 0, LTE category 1, and LTE category M. While LTE categories 0 and 1 are the first generation, LTE-M is the second generation and has improved cost and power consumption characteristics over the previous generation. LTE-M devices can communicate directly with a cellular network without the use of gateways [19]. However, one disadvantage of LTE-M is that each LTE-M device requires a SIM card.

Narrowband-IoT (NB-IoT): Narrowband-IoT (NB-IoT) is also known as long-term evolution (LTE) Cat NB1 and was developed by 3GPP for long-range communications at a low bit rate for public network cellular operators [20]. NB-IoT is the third generation of IoT radio protocols developed by 3GPP. The use of the LTE protocol stack provides a reduction in device costs and battery consumption since LTE is widely deployed worldwide by cellular operators for public mobile broadband service. Several features of the base LTE protocol are not supported. However, for example, handover, measurements to monitor the channel quality, carrier aggregation, and dual connectivity. In addition, NB-IoT does not require the deployment of a dedicated gateway, unlike many other IoT wireless technologies, since it runs over public operator networks.

5.4   IoT Sensors and Devices

IoT sensors and two-way communications have been integrated into the electric power grid network in generation, transmission, distribution, and customer domains as part of the transition to smart grid systems. Since most IoT sensors are deployed in the distribution and customer domains, this section gives examples of different smart devices in these domains, as shown in Fig. 5.3.

5.4.1   Smart Devices in the Distribution Domain

The distribution domain starts from a distribution substation and ends at customer premises. Smart devices in the distribution domain may include voltage regulators, distributed energy resources (DER), line sensors, switches, sectionalizer/reclosers, capacitor banks, distribution transformers, smart meters, and load control switches.

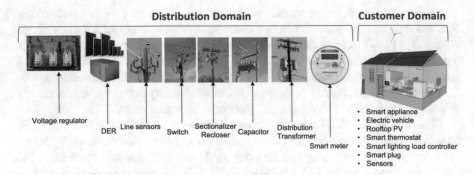

Fig. 5.3 Smart devices in the distribution and customer domains

DER: Distributed energy resources (DERs) have become popular for use at the distribution level. These may include community-scale battery energy storage, distribution-level solar PV, and diesel generators. A smart inverter may be deployed together with a distribution-level solar PV to allow an electric utility to remotely monitor and control the operation of the DER, enabling the adjustment of distribution-level voltage and active/reactive power control.

Voltage regulator: The voltage regulator senses the distribution-level voltage and automatically adjusts the voltage to provide a consistent voltage level. A typical voltage regulator can provide a regulation capability of +/−10% of the system voltage using 32 steps. With the high penetration of solar PV, the fluctuation of solar PV output may result in frequent changes in regulation steps of voltage regulators.

Line sensor: line sensors include both voltage sensors (potential transformers) and current sensors. Line sensors provide the overall distribution-level visibility to a control center, enabling identification of overcurrent/voltage conditions, phase-loss conditions, and detecting below normal voltage, that is, brownout conditions.

Disconnect switch: Disconnect switches are used to open a distribution circuit or disconnect electrical equipment from a distribution circuit. They typically operate at little or no-load current to prevent a high-voltage arcing problem. With two-way communications, the status of a disconnect switch can be controlled remotely from a control center.

Sectionalizer/recloser: Distribution sectionalizers and reclosers are load break switches that can disconnect a circuit carrying a load current. These devices break the circuit in the presence of a fault, a short circuit to the ground. These devices can communicate with a control center using a variety of communication protocols, like DNP3.0, IEC61850, Modbus, etc. Newer sectionalizers/reclosers can be networked together to enable fault detection, fault isolation, and circuit restoration collaboratively.

Capacitor bank: Capacitor banks normally include several capacitors of the same rating. In a smart grid, capacitor banks play an active role in Volt/VAR control applications where they are used to provide voltage support and power factor correction in a distribution circuit.

Distribution transformer: Distribution transformers step down the voltage from the level used in distribution lines to the level suitable for use at customer premises. Smart distribution transformers incorporate the ability to monitor their loading level (e.g., voltage, current, frequency, harmonics, power consumption, and power factor), together with their operating conditions, such as temperature and oil level, and enable remote data monitoring and collection for further analysis.

Smart meter: A smart meter is a connection point between a distribution domain and a customer domain. It measures the electrical energy consumption of a customer supplied by a distribution line. A typical smart meter enables data collection at 15-min intervals. With a smart meter, a customer may retrieve real-time and historical energy consumption information, as well as electricity price signals. Smart meters also enable an electric utility to gain distribution-wide voltage visibility at customer premises, as well as system-wide power outage status.

5.4.2 Smart Devices in the Customer Domain

The customer domain starts behind the meter and may include smart appliances, electric vehicles, rooftop solar PV, smart thermostats, smart lighting load controllers, smart plugs, and various types of sensors.

Smart appliance: Smart appliances include appliances that can shift their time of operation to avoid peak electricity prices. For example, a smart refrigerator can defer its defrost cycle, a smart dishwasher that can shift its start time, and a smart water heater change the time when it switches on to avoid operation at the peak electricity time of use price. These devices can actively participate in peak demand reduction and allow remote monitoring and control from the Internet.

Electric vehicle: Many electric vehicles (EVs) now have built-in communications that allow a user to monitor vehicles' battery levels via mobile applications. A smart EV may be able to communicate with a home energy management system to avoid charging at peak price hours or communicate with a smart transformer to avoid charging at peak transformer load hours.

Rooftop solar PV: Rooftop solar PVs as the cost of installing residential solar PV declines. Rooftop solar PV panels together with a smart inverter and two-way communications allow monitoring and control of PV power output. An electric utility can remotely monitor and control distributed solar PV panels via smart inverters to prevent overvoltage at customer premises.

Smart thermostat: A smart thermostat reports on indoor temperature and humidity and accepts control commands sent via the Internet. Some smart thermostats have a built-in occupancy sensor, which can detect the presence of a person in a space or send a reminder when it is time to change an air filter. Smart thermostats utilize different communication protocols, including Wi-Fi, ZigBee, BACnet, and Modbus.

Smart lighting load controller: A smart lighting load controller can control a light fixture on or off state or dim the light based on visual to reduce energy consumption. Newer light fixtures can change the color of the light and provide data for monitoring. In addition to the control of individual light fixtures, a controller can be installed to allow lighting control by zone. The higher granularity of zones in space offers greater control flexibility. A smart lighting load controller can communicate via a variety of protocols, including Wi-Fi, ZigBee, BACnet, and Digital Addressable Lighting Interface (DALI).

Smart plug: A smart plug is an outlet that plugs into another outlet on one side and an appliance on the other to enable control of the attached appliance. The appliance can be anything, such as a refrigerator, a fan, or a lamp. A smart plug can accept a control command sent via the Internet to switch the attached load on or off. Some smart plugs have a built-in schedule where the status of an attached device can be set in advance. With a smart plug, it is similar to transforming a regular appliance into a smart appliance having a remote control feature. A smart plug can also be integrated with IFTTT (IF This Then That) rule-based system to enable control based on an event. For example, a floor lamp can be turned on via a smart plug at sunset and turned off at sunrise.

Sensor: Many types of sensors are available to detect environmental conditions, such as those that measure temperature, humidity, and ambient light, and report the data to applications on the Internet. Air quality sensors are available that can monitor carbon dioxide or PM2.5 levels. Occupancy sensors can help save energy by connecting them to smart thermostats, lighting load controllers, and smart plugs, enabling the devices to switch on and of depending on whether someone is in the room. These IoT sensors enable a user to monitor real-time information gathered by the sensors and analyze historical data for trend analysis and other analytics.

5.5   An Energy Internet of Things (Energy IoT) Use Case: Electric Vehicle (EV) Charging

While smart grid systems, applications, and services are primarily concerned with providing utility tools for managing grid assets strictly on the DSO side of the meter, Energy IoT is more concerned with managing and providing information on behind-the-meter assets to both the DSO and the customer. In this section, an Energy IoT use case, electric vehicle (EV) charging, is examined in detail. EV charging is a good example of how an Energy IoT application can foster the development of a business ecosystem, provide a valuable service to consumers, and contribute to societal goals through encouraging widespread adoption of products and services that generate less carbon pollution, thereby contributing less to global warming.

A system architecture and protocol standards for EV charging have existed since 2015, but the architecture and protocols are still immature. On many of the interfaces, there are multiple protocol "standards" for the same interface with each protocol "standard" being promoted by a separate charging clearing house network or international standards organization. In addition, while charging EVs could act as an excellent source of flexible load for the grid, relatively few charging network operators communicate with their DSOs. While the actual shape of a future, fully interoperable internationally standardized EV charging network is still unclear, in the following sections, the architecture and protocols available today that are likely to influence the standardized, interoperable EV charging network of tomorrow are reviewed.

5.5.1   EV Charging Network Functional Architecture

The EV charging network functional architecture is shown in Fig. 5.4. The overall EV charging network can be broken down into the following subnetworks and their functional entities:

- The EV Charging Access Network connects EVs to Electric Vehicle Charging Stations (EVCSes) and EVCSes to the Charge Point Operator (CPO) that owns and maintains them, either directly or through a local controller (not shown). The EVCS is the point where the EV obtains electricity. The CPO handles the EVCS IT interface, physically maintains their EVCSes, and arranges for the electricity required by customers to charge their EVs. The CPO connects into the Energy Management Network for electricity and separately through a Clearing House to the Customer Access and Roaming Settlement Network for billing customers and settling roaming between eMobility Service Providers (eMSPs).
- The Energy Management Network connects the CPO to the building or site Energy Management System (EMS) if one is available and/or to the DSO, so the

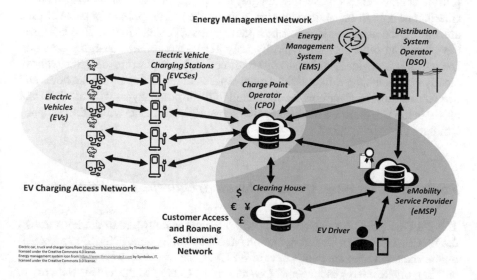

Fig. 5.4 Charging network functional architecture

CPO can schedule loads and, if the charging station supports vehicle-to-grid (V2G), provide power to the grid. Such scheduling of loads and energy provisioning is called "smart charging."

- The Customer Access and Roaming Settlement Network handles two functions:

   - On the Customer Access side, connecting EV drivers to their home eMSP, where the drivers have their accounts, through a smartphone app. The eMSPs primarily serve the EV drivers and handle the customer interface.
   - On the Roaming Settlement side, connecting serving CPOs that an EV driver visits with the EV driver's home eMSP through a Clearing House or directly if the CPO and eMSP are affiliated. The Clearing House enables roaming settlement when an EV driver uses a charging station owned by a CPO that is not affiliated with his home eMSP. The Clearing House arranges for usage records [called Charge Detail Records (CDRs)], including kWh and cost for a charging session to flow from the serving CPO back to the home eMSP, and for payment to flow in the opposite direction.

There are some regional differences between countries, particularly in the Customer Access and Roaming Settlement Network. For example, in the US, most CPOs are also eMSPs, EV drivers use RFID cards or apps on their smartphones provided by their eMSPs/CPOs to authenticate and authorize charging, and there is no roaming enabled between CPOs. In Europe, there are three major charge roaming networks (Hubject [21], Gireve [22], and e-clearing.net [23]) and a few minor ones. EV drivers can choose from many eMSPs to handle their account and can visit chargers from different CPOs by convenience. Each eMSP and CPO belongs to one or more charge roaming networks; roaming between the networks is enabled, and in some cases, EV drivers can use Plug and Charge (part of the ISO 15118 standard [20]) to charge without presenting any physical authentication like an RFID card or smartphone app. In Japan, the situation is more like in Europe, except some eMSPs are run by the automobile companies themselves (Nissan, Toyota, Honda, Tesla, etc.), while in other regions, typically only Tesla runs its own charging network. Despite these regional differences, the functional architecture described above clearly represents the different functional entities in the EV charging network, and even if two functional entities are combined into one company, as with the CPO and eMSP in the US, there are likely to be two separate departments within the company handling the two functions.

5.5.2   EV Charging Access Network Architecture and Protocols

The EV Charging Access Network is where power is delivered to the EV through the EVCS. The CPO's Charging Station Management System (CSMS) is responsible for managing the operator's fleet of EVCSes, coordinating with the DSO or the building or site energy management system in the Energy Management Network to enable smart charging, and coordinating with entities in the Customer Access and

Roaming Settlement Network to enable authorization, settle bills, and roaming. The EV itself can be a light-duty vehicle, like an electric car, or a medium or heavy-duty vehicle, like a truck or heavy off-road equipment like an electric backhoe, or it could even be part of a fleet run by a company. Micro-mobility solutions like electric bikes, electric skateboards, and electric scooters are out of scope. There are no charging station standards yet for electric bikes and skateboards since they are mostly charged from standard domestic outlets. While ISO has recently started a standardization committee for electric scooters (ISO TC 125 [24]), it has yet to issue any specifications for charging stations.

5.5.2.1   EVCS to EV Interface

The EVCS to EV interface is where the EV connects to the EVCS to charge. A variety of hardware standards apply to the EVCS charging connector hardware depending on the type of power (AC or DC, voltage, etc.), vintage, and geographic location where the charging station is deployed. Initially, chargers were either a Type 1 or Type 2 for AC charging or a Type 3 for DC fast charging. Type 1 and Type 2 AC chargers are covered by the SAE J1772 standard [25], commonly called the "J Plug" because of its shape. Type 3 DC fast chargers include the Combined Charging System, which allows both AC and DC charging from the same connector (Combo 1 and Combo 2) [30] and the Chademo charger [27], which allows DC charging only, developed by Nissan originally for the Nissan Leaf and now on other Nissan EVs. Charging connectors support the following protocols:

- The original 2001 SAE J1172 analog protocol [28] is used in older Type 1 and Type 2 charging stations. The charging station and the EV communicate by measuring voltage drops across a low voltage line. No complicated electronics or digital protocol is involved. The intent was to keep the charging station equipment simple and robust since the standard was developed prior to extensive microminiaturization of robust electronic components for smartphones that are built into newer charging stations. This protocol was subsequently incorporated into the IEC 61851-1 standard [29].
- IEC 61851-24 [34] is a standard for DC fast charging signaling adopted by the Chademo Association [30] in 2014 and implemented in Nissan EVs. IEC 61851-24 includes V2G support.
- ISO 15118 is a collection of standards for layers 1 through 7 of the network stack governing communication between an EV and an EVCS [20]. Specific subparts include:

    - ISO 15118-3 specifies the HomePlug GreenPHY [31] power line communication protocol for layers 1 and 2 (physical and link layers).
    - ISO 15118-8 specifies Wi-Fi (IEEE 802.11n) [32] for layers 1 and 2.
    - ISO 15118-2 specifies protocols from the Internet protocol suite (IP, UDP, TCP, TLS, DHCP, etc.) for layers 3 and 4 (network and transport layers), and various application-specific protocols for layers 5–7. Among the application layer pro-

tocols are the Plug and Charge security protocol allowing an EV driver who has registered to charge without showing any credentials to the EVCS, for example, no RFID card or smartphone app required. While part of the Plug and Charge protocol is executed between the EVCS and EV, most of the protocol concerns authorization and settlement and will be discussed in Section 5.5.5.2.

– ISO 15118-20 is an updated second-generation protocol expected to be published in 2021 that will include V2G signaling.

As an example of a protocol for communication on the EV to EVCS interface, Fig. 5.5 illustrates the IEC 61851-24 protocol to initialize DC fast charging in a vehicle equipped with a Chademo fast charging port [33]. The vehicle and charging station must first exchange information on the capabilities of both sides. For the vehicle, this includes the maximum charge voltage, battery parameters, maximum charge time, etc. For the charger, this includes the charging voltage and current, error parameters, etc. Once both sides have checked for compatibility, the charging session can be started.

5.5.2.2   EVCS to CSMS Interface

In contrast with all the other interfaces in the EV charging network architecture, the only communication protocol standardized on the EVCS to CSMS interface is the Open Charge Point Protocol (OCPP) [38]. OCPP 1.6 is in widespread implementation by EV charging station manufacturers and CSOs in the US and Europe, and in 2020 version 2.0.1 was released. Version 2.0.1 supports ISO 15118 on the EV to EVCS interface, including Plug and Charge. OCPP was originally developed as an industry standard by the Open Charge Point Alliance (OCPA) [35], and as of version 2.0 has been taken over by OASIS, an official standardization organization [36].

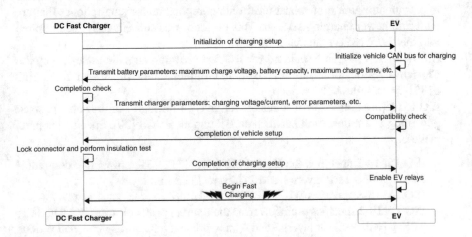

Fig. 5.5 IEC 61851-24 protocol for initialization of DC fast charging

OCPP runs over the Internet protocol stack between the EVCS and the CSMS, or to a local controller that connects to the CSMS, either over the wide area if two sides are some distance from each other or over a corporate LAN if the two are within a corporate network. Version 1.6 was originally specified to use the SOAP protocol [37], which is defined in XML and runs over HTTPS, but in Version 2.0, the OASIS standardization committee updated the specification by adding a JSON [38] over Websockets profile [39], since JSON is much simpler for developers to read. Websockets run over the TCP transport layer.

The OCPP specification includes three security profiles: no security, server certificate only, and both client and server certificates, the latter two profiles running on TLS. TLS is required unless the network between the EVCS and CSMS is trusted, for example, it is running over VPN. In most cases, a standard Web server/e-commerce unidirectional authentication procedure is followed, with the server presenting a certificate and the client verifying it. The CSMS takes the role of the server and provides an X.509 server certificate to the EVCS, while the EVCS takes the role of the client and validates the certificate. However, the protocol also allows for mutual authentication, in which case the EVCS is provisioned with and presents an X.509 client certificate to the CSMS in addition.

OCPP supports messages for the following functions:

- EVCS service life cycle management, including provisioning, firmware upgrade, and diagnostics,
- Installation and renewal of certificates for the ISO 15118 Plug and Charge protocol,
- Authorization of an EV driver to unplug the connector from a vehicle,
- Transaction initiation and termination, and meter values to report on the progress of charging both within and outside of a transaction,
- Providing the driver with information on the tariff through the EVCS display,
- Informing the CSMS about EVCS availability and allowing the CSMS to reserve a charging station on behalf of a specific EV driver,
- Display a message on the EVCS display,
- Remotely starting and stopping a transaction, unlocking a connector, or triggering other messages in the protocol,
- Smart charging, both with and without ISO 15118 on the EVCS to EV interface.

The protocol also includes an extensibility feature allowing messages not part of OCPP to be transferred between the CSMS and the EVCS.

5.5.3   Energy Management Network Architecture and Protocols

Communication between the Energy Management Network functional entities utilizes one of the following three protocol standards:

- The Open Automated Demand Response (OpenADR) protocol is an industry-standard protocol originally developed by Lawrence Berkeley National Laboratory but now maintained by an industry-standard group called the OpenADR Alliance [40] and first released in 2009. The latest version is 2.0 released in 2013, and includes two different profiles 2.0a and 2.0b, with the 2.0a profile being a simplified version of the protocol limited to only the message exchanges required to signal an event. OpenADR runs over two different transports: HTML over TLS and XMPP [41], a message bus protocol. The application layer protocol in both cases uses XML for data formatting.
- IEEE 2030.5 [43] was first released in 2018. It enables a DSO to communicate with DERs, including behind-the-meter assets for load control, time of day pricing, management of distributed generation, electric vehicle charging, etc. The transport layer protocol is HTTP/HTTPS, and the application layer protocol is defined according to the REST architecture. The 2030.5 protocol covers a broad variety of DERs, everything from flexible load devices such as electric hot water heaters to solar inverters, and includes profiles for over 30 different applications, EV charging among them.
- The Open Smart Charging Protocol (OSCP) [46] was originally developed by the Open Charge Alliance for the specific purpose of communicating a 24-hour forecast of available grid capacity from the DSO to the CPO's CSMS, so that the CSMS could schedule charging loads on the EVCS network to minimize grid impact and demand charges. Version 1.0 released in 2015 was designed specifically for communicating between the CSMS and the CPO's DSO for smart charging. In 2020, Version 2.0 was released with a considerably expanded scope. The protocol was generalized to apply to any DER connected to any electricity provider, and the architecture and entity names were correspondingly generalized away from the original purpose of OSCP 1.0. The transport layer protocol for OSCP is HTTP/HTTPS, and the application layer is designed based on the REST architecture with JSON used for data formatting.

These protocols allow the CSMS to obtain information about electricity availability and schedule charging loads from the electrical network supplying the EVCSes. The ability to schedule and finely control the transfer of energy in this fashion opens up a broad variety of applications that can help to stabilize the grid or reduce a building's or a site's utility billing charges. For example, if the CSMS is connected to a building or site EMS, scheduling of loads can help the building owner avoid large demand charges on their utility bill or can enable charging when electricity is especially cheap due to abundant renewable power generated from the building's on-site solar panels. On the CSMS to DSO interface, scheduling can reduce charging loads when the grid is experiencing a peak load event, helping the DSO mitigate peak loads and avoid rolling blackouts if the available supply from the TSO is less than the load.

The ability to control demand in this fashion has traditionally been called demand response. "Demand response" means actions taken to reduce the load on the grid when supply threatens to fall short of demand and covers a subset of the actions included under the more broadly scoped term "flexible loads," which includes loads that can additionally ramp up if electricity is especially abundant. If the EVCSes and connected EVs support V2G functionality, the CSMS can also schedule transfers of electricity in the opposite direction, from the EVs to the electrical network. V2G electricity transfers can help DSOs avoid peak events by providing additional supply when the supply from the TSO falls short.

5.5.3.1   EMS to CSMS and EMS to DSO Interfaces

If the building or site has an EMS, the OCPP 2.0.1 specification [34] proposes two alternatives to the architecture, shown in Fig. 5.6. Both depend on having a Local Controller that handles the communication as an intermediary between the EVCSes and the CSMS. In the first alternative, shown to the left at (a) in Fig. 5.6, the EMS connects to the Local Controller, and the Local Controller handles resolving the scheduling demands from the CSMS and EMS. In the second alternative, shown to the right at (b) in Fig. 5.6, the EVCSes are connected to both the Local Controller and to the EMS.

In (a), the EMS obtains the grid scheduling from the DSO through OpenADR or IEEE 2030.5, communicates it to the Local Controller using IEEE 2030.5, and the Local Controller handles the charging scheduling on the EVCSes using the smart charging protocol in OCPP. In (b) the EMS communicates directly with the EVCSes using IEEE 2030.5. If there is a conflict between the charging schedule sent by the CSMS and the EMS, the Local Controller needs to communicate that back to the CSMS. An example is if the building owner prioritizes HVAC loads over EV charging. With these architectures, the CSMS is not required to connect to the DSO directly since the EMS handles the DSO connection.

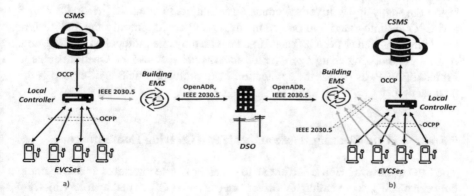

Fig. 5.6 Alternative charging access network architectures

5.5.4   Customer Access and Roaming Settlement Network Architecture and Protocols

The Customer Access and Roaming Settlement Network is the most complicated of the three subnetworks because there are multiple functions. These functions can often be combined into one entity. The Clearing Houses enable roaming between CPOs/eMSPs by acting as an intermediary when a visiting EV driver with an account at their home eMSP charges at a serving CPO affiliated with another eMSP. In addition, the Customer Access and Roaming Settlement Network allows EV drivers to access their eMSP account through a smartphone app. There are four interfaces in this subnetwork: the eMSP to EV Driver interface through the smartphone app, the CPO to eMSP interface, the CPO to Clearing House interface, and the eMSP to Clearing House interface. These interfaces are described in the following subsections.

5.5.4.1   eMSP to EV Driver Smart Phone App Interface

The eMSP to EV Driver smartphone app interface allows the driver to check on the status of their account and perform other functions. The app supports functions for locating charging stations, joining a station waiting list, reserving a station, starting and stopping charging, reporting on the progress of a charging session, and reporting on the amount of credit in the account or paying for charging sessions. The eMSP can, in principle, use any transport layer protocol and structure the application layer protocol in any way it chooses since it will be implementing both the client-side in the app and the server-side in the eMSP's Web server, but HTTP/HTTPS is a popular choice.

For the application layer protocol, OCPP can also serve between the eMSP and the smartphone app since it supports functionality for remotely controlling an EVCS and an extension feature that allows other messages to be exchanged between both sides that are not part of the base protocol. The extension feature could be used to report the status of the driver's account, for example. In countries where the eMSP and CPO are integrated into one company, OCPP could simplify the architecture since the combined eMSP/CPO need to deal with only one protocol on the back end. Apps for authorizing charging in the Intercharge network use the Open Intercharge Protocol (OICP) as part of the application layer protocol. OICP is discussed in the next section.

5.5.4.2   CPO to Clearing House and eMSP to Clearing House Interface

The CPO to Clearing House and eMSP to Clearing House interfaces enable roaming between eMSPs, so a visiting EV driver can use an EVCS owned by a serving CPO not affiliated with their home eMSP. There are three protocols used on these interfaces, roughly aligned with the three European charge roaming networks:

- The Open InterCharge Protocol (OICP) [43] was developed by Hubject [21], a joint venture between several German companies: the automobile companies BMW, Daimler, and Volkswagen, the electrical and automotive equipment suppliers Bosh and Siemens, and the DSOs Energie Baden-Württemberg AG and Innogy/E-ON. The Intercharge network is the largest in Europe and includes over 40,000 charge points as of December 2019 in 17 European countries, Japan, Israel, and New Zealand. Development on OICP was started in 2012, and version 2.3 was released in 2020. Beginning with the 2.3 release, the protocol is divided into two parts that are versioned independently, one part for the CPO to Clearing House interface and the other for the eMSP to Clearing House interface. The protocol is designed according to the REST OpenAPI 2.0 standard [44], which uses JSON for data formatting in the application layer protocol and HTTP/HTTPS for the transport layer protocol.
- The eMobility Interoperation Protocol (eMIP) [45] was developed by the French clearing house Gireve [22] and had over 26,000 charging points throughout the EU as of January 2019. The latest version is 1.0.13 released in 2020. Unlike the other protocols in the EV charging ecosystem, eMIP remains the intellectual property of Gireve, and while the specifications have been openly published and a license is available without charge, Gireve maintains exclusive rights to modify the protocol. The eMIP application layer protocol uses SOAP version 1.2 for specifying data formats and HTTPS as the transport layer protocol.
- The Open Clearing House Protocol (OCHP) [47] was developed and is maintained by the e-clearing.net [23] Clearing House. The e-clearing.net network was launched in 2014 as a joint venture between the Dutch foundation ElaadNL and the German company smartlab Innovations GmbH, and now also includes Blue Corner in Belgium and ladenetz.de in Germany as well. As of December 2020, e-clearing.net has 18 member organizations and 9500 charging stations primarily in Belgium, the Netherlands, and Germany. The latest version of OCHP is version 1.4, released in 2016. The application layer protocol uses SOAP version 1.6 for specifying data formats and runs over the HTTP/HTTPS transport layer protocol. OCHP also has a version for communicating directly between a CPO and eMSP without going through a Clearing House, called OCHPdirect, which is briefly discussed in Section 5.5.5.3.

These protocols offer four primary functions on the two interfaces:

- Allow the serving CPO to authenticate a visiting EV driver or EV for charging and authorize a charging session with the visiting EV driver's home ESP via the Clearing House if necessary,
- Communicate a record of the charging session electricity usage and cost, or CDR, from the serving CPO via the Clearing House to the driver's home eMSP after the charging session has ended,
- Provide an EV driver with information on EVCS availability and reserve and EVCS in a nearby location,
- Allow bulk uploading of authorization data to the Clearing House.

In addition to charging sessions, OCHP offers some additional services, including navigation information for finding a charging station and the ability to manage and bill for parking spots.

Since the latter three functions involve a relatively straightforward exchange of formatted records, we focus on the authentication and authorization function, as it is the key to establishing a level of trust between the parties that is crucial for anchoring an EV charge roaming business ecosystem. The authentication and authorization function allows a serving CPO to ensure that the visiting EV driver has an account with an eMSP and that they will be compensated for the electricity consumed during a visiting EV driver's charging session. The Clearing House acts as a trusted intermediary and enables the connection between the serving CPO and home eMSP when the serving CPO is not affiliated with the EV driver's home eMSP. The protocols on the CPO to Clearing House and eMSP to Clearing House interfaces use two different methods for authentication and authorization: requiring the EV driver to present some type of physical credential and Plug and Charge, where the EV driver simply plugs in the EV and the EV authenticates itself.

When the EVCS requires the visiting EV driver to present a physical credential, all three protocols handle authentication through either an RFID card or a smartphone running an app that communicates with the EVCS through the smartphone's Near Field Communication device. The EVCS reads the visiting EV driver's account id and home eMSP id from the RFID card or smartphone app and communicates them to the serving CPO's CSMS. The serving CSMS appends the serving CPO's id and routes the three identifiers to the Clearing House, which uses the home eMSP id to route the request to the home eMSP. If the home eMSP can identify the driver as having an account in good standing, it returns an authorization token for the serving CPO to proceed with charging.

The Intercharge network additionally uses a different protocol based on the EV Driver scanning a QR code with their smartphone app. An EVCS in the Intercharge network has a QR-code affixed to their housing that allows the smartphone app to initiate the request for authorization rather than the EVCS. The visiting EV Driver scans the QR code on the EVCS, and the serving CPO's id, serving EVCS id, home eMSP's id, and visiting EV driver's account number is then sent from the app via OICP to the Intercharge Clearing House at Hubject, which routes the request for authorization to the EV driver's home eMSP. The home eMSP replies via the Clearing House back to the serving CPO, either authorizing or denying the charging session.

The Plug and Charge protocol allows an EV driver to charge without presenting any physical credential. As the name implies, the EV driver simply plugs the EV into the serving EVCS, and the serving EVCS authenticates and requests authorization from the EV driver's home eMSP via the serving CPO and the Clearing House. In effect, the EV itself acts as the physical credential. Plug and Charge was standardized as part of the ISO 15118-2 protocol [20]. Figure 5.7 illustrates the Plug and Charge certificate provisioning and session authorization protocols.

Plug and Charge uses X.509 public-key certificates issued through a complicated series of interlinking certificate authorities to authorize charging. The EV

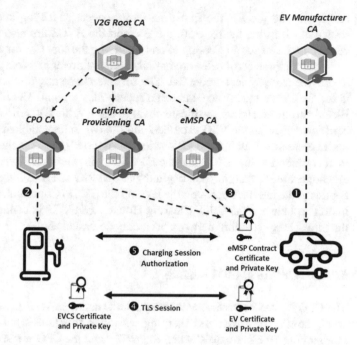

Certificate authority icon from https://www.entrust.com
Electric car, truck and charger icons from https://www.icons-icons.com by Timofei Rostilov
licensed under the Creative Commons 4.0 license.

Fig. 5.7 Plug and charge certificate provisioning and session authorization protocol

manufacturer runs a CA that issues certificates for vehicles, which it manufactures. The EV certificate contains the vehicle identification number (VIN). A vehicle to grid (V2G) CA acts as the root CA for the CAs run by the CPO and eMSP. In addition, a Certificate Provisioning CA under the V2G CA helps with the initial installation of a contract CA from the eMSP into the EV. A contract CA provides proof of an EV driver's account with the home eMS and contains the EV driver's identification. All public/private key pairs are generated with an elliptic curve cryptosystem, the elliptic curve Diffie–Hellman protocol is used for generating shared session keys, and the AES 128 bit shared key cryptosystem is used for encrypting session traffic.

Initially, the different entities that participate in a charging session need to be provisioned with certificates. When the EV is manufactured, the EV manufacturer installs the EV certificate and corresponding private key into hardware security module-like device on the car called the electric vehicle communication controller (EVCC), shown at (1) in the figure. On the CPO side, after installation, an EVCS is provisioned by the CPO with a certificate and private key identifying the EVCS into similar hardware security module-like device in the EVCS called the supply equipment communication controller (SECC), shown at (2). After the EV is purchased and the EV driver has signed up with an eMSP, an automated provisioning system assisted by the Certificate Provisioning CA installs the EV driver's contract certificate into the EV the first time they connect to a Plug and Charge EVCS, shown at (3) in the figure. Both the EV and the EVCS are also provisioned with certificate chains for all CAs up to and including the root CA for their certificates.

When all three certificates are in place, the EV driver can charge without having to provide any physical credential. The first step in setting up the charging session is for the EVCS and EV to establish a secure TLS session. The EVCS presents the EVCS certificate and its certificate chain to the EV, the EV validates it through the certificate chain to the V2G root CA, and the two sides establish a session key for the TLS session. Within the TLS session, after the EV presents the EV driver's contract certificate to the EVCS and the EVCS obtains authorization from the EV driver's home eMSP, charging can begin. Note that ISO 15118-2 Plug and Charge only applies to the interface between the EV and the EVCS. Communication on the other interfaces between the CPO, Clearing House, and eMSP is conducted with one of the three charge roaming network protocols described above.

5.5.4.3   CPO to eMSP Interface

The CPO to eMSP interface is where the CPO and the eMSP exchange information about the EV, EV driver, and charging session in situations where the EV driver is not roaming, in other words, when the eMSP and the CPO are affiliated. There are two protocol standards for this interface:

- The Open Charge Point Interface (OCPI) protocol, development of which was started in 2014 and was taken over by NKL [48] in 2015. NKL is a nonprofit in the Netherlands dedicated to promoting an open and accessible charging infrastructure. In 2020, the EVRoaming Foundation [49] was formed to guide further development. The latest version of the protocol is Version 2.2 [50]. The protocol uses HTTP/HTTPS for transport, JSON for data formatting and is designed based on the REST API architecture.
- The OCHPdirect, a variant on the OCHP protocol from Section 5.5.5.2 for direct communication between the CPO and eMSP. OCHPdirect uses SOAP version 1.1 over HTTP with TLS version 1.2 and server-side authentication only.

OCPI supports the following functions:

- The Versions function that asks for and replies to requests for starting a session. This allows both sides to negotiate the version of the protocol to be used.
- The Credentials function to generate tokens for use in authenticating HTTP messages. OCPI uses HTTP Token Authorization headers with a generated token.
- The Location function allows the eMSP to query for and the CPO to provide information on EVCS locations. The returned records have detailed information about the EVCS equipment, such as the power, the hours they are available, etc.
- The CDR function describes a concluded charging session so the eMSP and CPO can settle the billing.

- The Tariffs function provides information to the eMSP on the tariffs used by the CPO.
- The Tokens function allows the eMSP to push or the CPO to pull charging tokens for EV drivers in order to allow the CPO to directly authorize a charging session without consulting the eMSP.
- The Commands function allows remote commands to be sent to the EVCS to reserve or unreserve a charging station, to start or stop a charging session, or to unlock a connector.
- The Smart Charging function allows the eMSP and CPO to coordinate smart charging.
- The Hub Connection function provides information to the CPO about which eMSP servers are currently up and online.

OCHPdirect has a more limited set of functions that includes an additional extension protocol to OCHP. The extension allows an eMSP or CPO to discover the server endpoints for roaming partners from the Clearing House. In addition, the OCHP extension protocol allows the CPO to download authorization tokens from the Clearing House for identifying EV drivers from the eMSP partners. The tokens are valid for a period of 1 day and must be refreshed on a daily basis. This allows the CPO to authorize a charging session for a visiting EV driver directly. CDRs are then sent directly from the CPO to the eMSP rather than through the Clearing House.

5.5.5 Discussion

The architecture of the EV charging network outlined in this section is in many ways like the early Internet or the early cell phone network before the advent of agreed-upon standards. Most of the protocols defined on the interfaces are industry-standard protocols and are not maintained by any neutral standardization group. While industry-standard protocols can help to achieve technical interoperability in the early years of a developing network like EV charging, problems occur when competing economic interests in the business ecosystem growing around the network promote competing industry standards on the same interface in order to lock in or exclude companies. The EV charging network roaming settlement interface between the Clearing House, eMSP, and CPO is an example of such a development.

There have also recently been some discussion of security problems with Plug and Charge [51]. The problems stem not from the basic certificate validation procedure that is the root of the protocol's function, which is to make EV charging completely seamless, but rather from the way the EV driver's contact certificate is provisioned onto the vehicle. Analysis indicated that the certificate provisioning protocol might result in the introduction of a vulnerability allowing a "man in the middle" attack. In 2020 ISO launched ISO 15118-20 to address this and other concerns about the protocol, but the security concerns were not raised about the communication between the EVCS and the vehicle, which is all that ISO 15118 really specifies. They were raised about the supporting protocol infrastructure, and business concerns play into the development of that infrastructure.

Time will tell if the EV charging network architecture develops the way the Internet and cell phone architectures did, toward vendor and operator neutral standards where all players realized that everybody had a stake in ensuring interoperability. Typically, such standards can only be ensured by having a neutral standardization body, like IETF for the Internet or 3GPP for the cellular network, overseeing the architecture and protocol design, where everybody with a stake in the business ecosystem has a voice. Such a body has yet to develop for the EV charging network.

5.6   Summary

The Internet of Things (IoT) is a transformation of Machine-to-Machine (M2M) technology to meet the interconnection of intelligent devices and management platforms along with advanced communication and communication technologies. This chapter provides a comprehensive review of the Energy Internet of Things (EIoT) concept in the smart grid environment. This chapter also provides details on IoT communication and networking technologies and smart grid IoT applications. The discussed communication technologies include Wi-Fi Direct, Bluetooth Low Energy (BLE), ZigBee, Z-Wave, 6LoWPA, EnOcean, Thread, ANT+, Near-Field Communication (NFC), LoRa, SigFox, Narrowband-IoT (NB-IoT), and LTE-M. Such communication technologies supporting long-range, low-power (LoRa, SigFox, Narrowband-IoT (NB-IoT), and LTE-M) are leading in IoT applications. Those technologies have enormous potential as key enablers for IoT today due to their ubiquitous connectivity for smart grid applications in distribution and customer domains. The main challenges of communication technologies supporting IoT applications are the requirements to be secure, flexible, low-power, and easy to provision, manage, and scale while delivering robustness and acceptable latencies in performance.

Connectivity is a crucial success factor for IoT applications. It is expected a high performance in terms of reliability and availability. Trending IoT applications focus on cellular technologies, such as the 4G-based LTE-M and NB-IoT technologies, due to requiring global coverage and mobility. In addition, existing IoT applications rely on low-power WAN (LPWNA) technologies, such as LoRaWAN or Sigfox, as well as short-range or midrange wireless technologies, such as Wi-Fi, ZigBee, Z-Wave, Thread, Bluetooth Low Energy (BLE), 6LoWPA, EnOcean, and others.

An in-depth study of a specific Energy IoT use case, namely electric vehicle charging, revealed common features of Energy IoT applications in general and showed how Energy IoT can provide a valuable service to consumers and utilities that would otherwise be difficult to implement. In an Energy IoT application, typically, there is an end device or collection of devices, which communicates with an application server. The services provided by the application server are not available from the devices alone. In the case of EV charging, the devices are the EV itself and the charging station. The main service provided for the EV by the application server and EV charger is the ability to charge "on the go," away from the owner's home charger. The service provided for the charging station is energy management, so bulk EV charging electricity use can be controlled to avoid high billing costs, negative grid impacts, and better match use of renewable energy to supply. The application server also coordinates the use of the charging stations between EV drivers.

Another aspect of Energy IoT applications is the billing system for services. The billing system provides the EV drivers with simplified billing, and also, some EV charging systems allow owners to roam between different charging networks. Many Energy IoT applications come with a smartphone app that allows the user to connect up to the application server and configure it with their preferences. For EV charging, this allows EV drivers to see which chargers are unoccupied, to reserve chargers, and to turn charging off and on remotely. Finally, Energy IoT applications have a common thread of security as the services are tempting targets for hackers. Security on user facing services such as Energy IoT applications needs to be as good as security on backend systems run by the distribution system operator. Overall, Energy IoT applications have a large potential to provide useful services to consumers while helping the distribution system operators to optimize their networks and improve the use of renewable energy.

References

1. Difference between IoT and M2M, [Online]: https://www.geeksforgeeks.org/difference-between-iot-and-m2m/ (Accessed 2020-10-01).
2. Oracle, What Is the Internet of Things (IoT)? [Online]: https://www.oracle.com/internet-of-things/what-is-iot/ (Accessed 2020-11-01).
3. Josh Fruhlinger, What is IoT? [Online]: The Internet of things explained, https://www.networkworld.com/article/3207535/what-is-iot-the-internet-of-things-explained.html (Accessed 2020-10-23).
4. Chantal Polsonetti, Know the Difference Between IoT and M2M, [Online]: https://www.automationworld.com/products/networks/blog/13312043/know-the-difference-between-iot-and-m2m (Accessed 2020-10-23).
5. Calum McClelland, The Industrial Internet of Things - What's the Difference Between IoT and IIoT? [Online]: https://www.leverege.com/blogpost/difference-between-iot-and-iiot (Accessed 2020-09-01).
6. i-Scoop, Business guide to Industrial IoT (Industrial Internet of Things) [Online]: https://www.i-scoop.eu/internet-of-things-guide/industrial-internet-things-iiot-saving-costs-innovation/ (Accessed 2020-11-01).
7. Internet of Things – Software, [Online]: https://www.tutorialspoint.com/internet_of_things/internet_of_things_software.htm#:~:text=IoT%20software%20addresses%20its%20key,extension%20within%20the%20IoT%20network (Accessed 2020-08-01).
8. Rohde-Schwarz, Internet of Things (IoT) Testing, [Online]: https://www.rohde-schwarz.com/us/solutions/test-and-measurement/wireless-communication/iot-m2m (Accessed 2020-11-10).
9. What is Wi-Fi Direct? Here's everything you need to know, [Online]: https://www.digitaltrends.com/computing/what-is-wi-fi-direct/ (Accessed 2020-10-01).

10. Bluetooth Vs. Bluetooth Low Energy: What's The Difference? [Online]: https://www.link-labs.com/blog/bluetooth-vs-bluetooth-low-energy (Accessed 2020-08-01).
11. D. Bian, M. Kuzlu, M. Pipattanasomporn and S. Rahman, "Assessment of communication technologies for a home energy management system," ISGT 2014, Washington, DC, USA, 2014, pp. 1-5, https://doi.org/10.1109/ISGT.2014.6816449.
12. Devasena, C.L., 2016. IPv6 low power wireless personal area network (6LoWPAN) for networking Internet of things (IoT)–analyzing its suitability for IoT. Indian Journal of Science and Technology, 9(30), pp.1-6.
13. Thread IoT Wireless Technology, [Online]: https://www.electronics-notes.com/articles/connectivity/ieee-802-15-4-wireless/thread-wireless-connectivity.php
14. What is ANT+? [Online]: https://www.rfwireless-world.com/Terminology/ANT-basics.html (Accessed 2020-08-05).
15. What is NFC? [Online]: https://www.verypossible.com/insights/nfc-and-iot-what-you-need-to-know (Accessed 2020-11-15).
16. E. Bingöl, M. Kuzlu and M. Pipattanasompom, "A LoRa-based Smart Streetlighting System for Smart Cities," 2019 7th International Istanbul Smart Grids and Cities Congress and Fair (ICSG), Istanbul, Turkey, 2019, pp. 66–70, doi: https://doi.org/10.1109/SGCF.2019.8782413.
17. SigFox for M2M & IoT, [Online]: https://www.electronics-notes.com/articles/connectivity/sigfox/what-is-sigfox-basics-m2m-iot.php (Accessed 2020-11-20).
18. 3GPP, Home Standardization of NB-IOT completed; 2019, [Online]: http://www.3gpp.org/news-events/3gpp-news (Accessed 2020-11-20).
19. LannerAmerica, LTE-M vs LoRa: Who Will Win The IoT Race? [Online]: Available from: https://www.lanner-america.com/blog/lte-m-vs-lora-will-win-iot-race/ (Accessed 2020-09-15).
20. V2G Clarity, "What is ISO 15118?". [Online]: https://v2g-clarity.com/knowledgebase/what-is-iso-15118/ (Accessed 2020-10-01).
21. Hubject, "Hubject – customer friendly electric mobility with eRoaming". [Online]: https://www.hubject.com/en/ (Accessed 2020-10-01).
22. Gireve, "Home-Gireve". [Online]: https://www.gireve.com/home (Accessed 2020-12-01).
23. e-clearing.net, "Europe's borderless charging network". [Online]: https://e-clearing.net/ (Accessed 2020-12-01).
24. ISO TC 125, "IEC TC 125 Dashboard". [Online]: https://www.iec.ch/dyn/www/f?p=103:7:0::::FSP_ORG_ID,FSP_LANG_ID:23165,25 (Accessed 2020-11-26).
25. SAE, "SAE Electric Vehicle Conductive Charge Coupler, SAE J1772", October 13, 2017. [Online]: https://www.sae.org/standards/content/j1772_201710/ (Accessed 2020-11-22).
26. CharIn, "CCS Implementation Guideline". [Online]: https://www.charinev.org/ccs-at-a-glance/ccs-implementation-guideline/ (Accessed 2020-11-22).
27. Chademo Association, "Chademo Association and Protocol". [Online]: https://www.chademo.com/wp2016/wp-content/uploads/2019/05/2019%20CHAdeMO_Brochure_web.pdf (Accessed 2020-11-22).
28. Wikipedia, "SAE J1772". [Online]: https://en.wikipedia.org/wiki/SAE_J1772 (Accessed 2020-11-23).
29. IEC, "IEC 61851-1:2010". [Online]: https://www.iecee.org/dyn/www/f?p=106:49:0:FSP_STD_ID:6029 (Accessed 2020-11-26).
30. IEC, "IEC 61851-24:2014". [Online]: https://www.iecee.org/dyn/www/f?p=106:49:0:FSP_STD_ID:6033 (Accessed 2020-11-23).
31. Zyren, J., "The HomePlug Green PHY specification & the in-home Smart Grid", 2011 IEEE International Conference on Consumer Electronics (ICCE), March, 2011.
32. IEEE, "802.11n-2009". [Online]: https://standards.ieee.org/standard/802_11n-2009.html (Accessed 2020-11-23).
33. Haida, T., "IEC/EN Standardization". [Online]: https://www.chademo.com/wp2016/wp-content/uploads/2014/10/IEC_standarization_update.pdf (Accessed 2020-11-26).
34. Open Charge Alliance, "Open Charge Point Protocol: Version 2.0.1", March 31, 2020.

35. Open Charge Alliance, "Open Charge Point Alliance: Global Platform for Open Protocols". [Online]: https://www.openchargealliance.org/ (Accessed 2020-11-26).
36. OASIS, "OASIS OCPP Electric Vehicle Charging Equipment Data Exchange TC", [Online]: https://www.oasis-open.org/committees/tc_home.php?wg_abbrev=ocpp (Accessed 2020-11-26).
37. W3C, "SOAP Specifications", [Online]: https://www.w3.org/TR/soap/ (Accessed 2020-11-26).
38. Json.org, "JSON". [Online]: https://www.json.org/json-en.html (Accessed 2020-12-03).
39. Fette, I., and Melnikov, A., "The WebSocket Protocol," RFC 6455, IETF, December, 2011.
40. OpenADR Alliance, "OpenADR Alliance" [Online]: https://www.openadr.org/ (Accessed 2020-11-27).
41. Saint-Adre, P., "Extensible Messaging and Presence Protocol (XMPP): Core", RFC 6120, Internet Engineering Task Force, March, 2011.
42. IEEE, "IEEE 2030.5 – IEEE Standard for Smart Energy Profile Application Protocol". [Online]: https://standards.ieee.org/standard/2030_5-2018.html (Accessed 2020-11-28).
43. Hubject, "hubject/oicp: Open intercharge Protocol". [Online]: https://github.com/hubject/oicp (Accessed 2020-12-05).
44. Swagger, "Open API Specification: Version 2.0". [Online]: https://swagger.io/specification/v2/ (Accessed 2020-12-05).
45. Gireve, "eMIP Protocol". [Online]: https://www.gireve.com/wp-content/uploads/2020/06/Gireve_Tech_eMIP-V0.7.4_ProtocolDescription_1.0.13-en.pdf (Accessed 2020-12-05).
46. Open Charge Alliance, "Open Smart Charging Protocol: Version 2.0", October 12, 2020.
47. OCHP, "Open Clearing House Protocol (OCHP): The free international standard for free e-mobility interoperability". [Online]: http://www.ochp.eu/ (Accessed 2020-12-05).
48. NKL, "The Netherlands Knowledge Platform for Charging Infrastructure". [Online]: https://www.nklnederland.com/ (Accessed 2020-12-04).
49. EVRoaming Foundation, "Open Charge Point Interface". [Online]: https://evroaming.org/ (Accessed 2020-12-04).
50. EVRoaming Foundation, "OCPI Version 2.2". [Online]: https://evroaming.org/app/uploads/2020/06/OCPI-2.2-d2.pdf (Accessed 2020-12-04).

Chapter 6
Foundations of Big Data, Machine Learning, and Artificial Intelligence and Explainable Artificial Intelligence

Industry 4.0 is a concept describing the digitalization wave for industry and economy, leading toward the ultimate goal of a fully digital economy. Big data (BD), machine learning (ML), and artificial intelligence (AI) are key enabling technologies for Industry 4.0. This chapter focuses on the foundations of BD, ML, and AI to provide guidance primarily for energy professionals who need an introduction to these technologies and the core thinking behind them. Section 6.1 describes the foundations of big data. Section 6.2 discusses techniques for building artificial intelligence and machine learning models, specifically supervised learning, unsupervised learning, dimensionality reduction, clustering in unsupervised learning, reinforcement learning, and ensemble learning. In Section 6.3, explainable artificial intelligence (XAI) is discussed. While other AI techniques often don't provide clear guidance on why a particular decision has been reached and therefore are sometimes difficult for human decision-makers to accept, XAI techniques are designed to provide human decision-makers with understandable reasons for their recommendations. Finally, Section 6.4 summarizes that chapter.

6.1 Foundations of Big Data

Big data (BD) is a data processing technology field that focuses on collecting, cleansing, storing, analyzing, extracting information from, and interpreting large data sets. With the Internet, Internet of Things (IoT), sensor networks, and similar technologies generating huge amounts of data from their respective operation domains. BD and BD analytics provide the means to translate the data into actionable intelligence. It is estimated that more than 59 zettabytes (ZB) of data were generated globally by the end of 2020. Table 6.1 demonstrates the units of data volume in various scales with abbreviations.

© The Author(s), under exclusive license to Springer Nature Switzerland AG 2021
U. Cali et al., *Digitalization of Power Markets and Systems Using Energy Informatics*, https://doi.org/10.1007/978-3-030-83301-5_6

Table 6.1 Data volume units, values, abbreviations, and sizes

Unit	Abbreviation	Capacity
Bit	b	1 or 0
Bytes	B	8 bits
Kilobyte	KB	1024 bytes
Megabyte	MB	1024 kilobytes
Gigabyte	GB	1024 megabytes
Terabyte	TB	1024 gigabytes
Petabyte	PB	1024 terabytes
Exabyte	EB	1024 petabytes
Zettabyte	ZB	1024 exabytes
Yottabyte	YB	1024 zettabytes

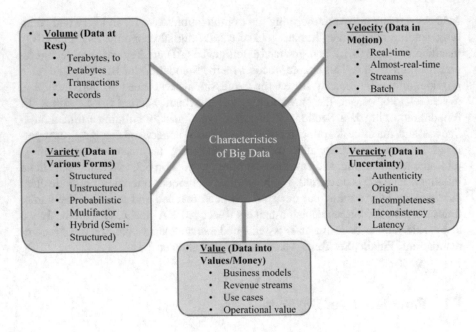

Fig. 6.1 Five key characteristics of big data

Such large amounts of data cause various challenges. These challenges include organizing adequate data storage media, scheduling computational tasks, ensuring data transaction speed with an acceptable quality of service for users, and similar issues [2]. However, BD provides the technology to translate this data in science and businesses such as health, media, energy systems, and defense into usable information. Unlocking BD's technical, social, and economic potential depends on the methodologies employed in addressing various technical challenges. Figure 6.1 shows five key characteristics of BD [1]:

- **Volume**: This characteristic treats the quantification of the data volume at rest. The units can be indicated in terms of bits to Yottabytes (YB). The projected data volume is set to increase dramatically in the coming years with new IoT devices, sensors, and similar technologies.
- **Velocity**: Velocity is the speed of the transactions generating and consuming data between various resources, processing, and storage domains. Data can be collected over time (batch processing) or streamed continuously in a real or almost real-time manner, depending on the application type.
- **Veracity**: Veracity is the characteristic that assures the data is accurate, valid, and clean (free of inaccurate or missing data points). Inconsistency, incompleteness, and lack of authenticity may increase the uncertainty of the data providence.
- **Variety**: Variety refers to the type or form of data that can be subcategorized as structured, unstructured, or semi-structured (hybrid).
- **Value**: Value is the combined characteristics of all 4Vs above and refers to the added value provided to the business unit or science domain in terms of revenue streams, operational values, etc.

BD analytic techniques are used to turn this massive tsunami of data into useful information and actionable insights. Commonly used BD analytic techniques are

- Statistical, such as regression algorithms,
- AI/ML algorithms,
- Database querying,
- Advanced data mining.

Visualization is an essential aspect of making big data more interpretable and understandable. The BD analytics algorithms like advanced AI/ML are employed in coordination with various visualization and graphical representation methods to satisfy the following capabilities:

- Demonstrating the data and results as clearly as possible,
- Processing multiple types of incoming data,
- Making the large amounts of data intelligible and coherent,
- Applying various filters to fine-tune the results,
- Providing high-quality exploratory data analytic features,
- Interacting with the data sets even when the data is in motion.

Common types of data visualization methods are

- Graphics and plots
- Tables
- Heatmaps
- Maps
- Charts
- Infographics
- Histograms
- Dashboards

BD is one of the leading enabling technologies of new and next-generation digital economies, and its use is spreading across various domains. It also accommodates a broad spectrum of use cases, for example,

- Sentiment analysis,
- Predictions,
- Segmentation,
- Fraud detection and prevention,
- Cyberphysical security and intelligence,
- Price determination and optimization.
- Energy sector use cases such as smart electric meter reading systems, forecasting algorithms, anomaly detection based on BD

Data science is the interdisciplinary domain that utilizes various scientific methodologies, algorithms, and frameworks to generate new knowledge and insights from big data, as illustrated in Fig. 6.2. It aims to extract knowledge from a data set in a systematic way.

A typical life cycle of a data science project includes the following stages:

- Data collection and discovery,
- Data preparation,
- Mathematical and statistical modeling,
- Algorithm development and programming,
- Visualization,
- Commutation and interpretation.

Fig. 6.2 Overview of data science

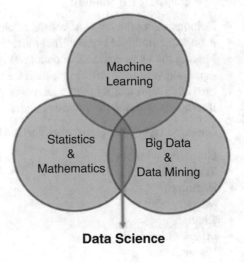

6.2   Artificial Intelligence and Machine Learning

Artificial intelligence (AI) is a broad term that covers the theory and application of computer systems to perform tasks that typically require human intelligence, such as speech and image recognition. In most cases, such tasks cannot be completed by algorithmic means. Machine learning (ML) is a subset of AI in which algorithms and statistical models are trained to analyze and draw conclusions from data patterns. Moreover, deep learning (DL) is a subset of ML that uses artificial neural networks to implement ML systems. These fields are closely related, as demonstrated in Fig. 6.3.

In Fig. 6.4, the overall ML/AI process is illustrated. The various steps in the process include the following:

Raw data: The entire ML/AI processing phase starts with raw data collection and storage. Once the raw is collected from various sources, for example, using sensors from the observation domain, generated by other agents such as sentimental outputs of the Internet users or syntactically generated data resources, it is stored. The data storage environment can be structured as databases in public cloud storage systems or dedicated servers. Typical databases are No SQL databases or comma-separated values (CSV) files.

Preprocessing: Preprocessing methods like normalization, data cleansing, feature extraction, and dimensionality reduction are among the commonly used techniques. Preprocessing makes raw data usable by the learning model algorithms. The aim of normalization is to transform and fit the values of raw data to a more

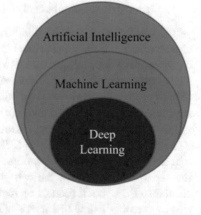

Fig. 6.3 Overview of AI, ML, and DL

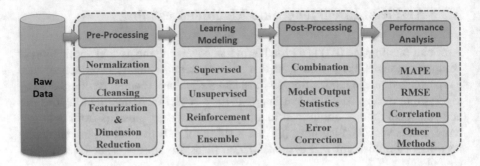

Fig. 6.4 Overview of the ML processes

useful format or a common scale without losing the main characteristics of the data. Data cleansing performs plausibility analysis to make sure that there are no missing values or implausible data points. Feature extraction splits data into a set of features to find essential characteristics that make one or another pattern recognizable. As a result, each data point is assigned a particular class. Dimensionality reduction is a key requirement for problems that contain large data sets and high-dimensional spaces. High dimensional means that the number of categories or dimensions on which the data could potentially be classified is extremely large, which makes computations using the data quite difficult. In ML, features are individual independent variables or characteristics of the observed data set that act as an input in the given system. The number of features can surpass the number of observations if the system has high-dimensional data. Various techniques are used to reduce dimensionality [4]:

- Principal component analysis (PCA)
- Linear discriminant analysis (LDA)
- General discriminant analysis (GDA)

PCA is the most popular dimensionality reduction method since it preserves the characteristics of the most critical information from the larger data set. Smaller and representative data sets are easier to gain insights from and explore and also require less computational power in comparison to processing larger data sets. The main idea of principal component analysis (PCA) is to reduce the dimensionality of a data set consisting of many variables correlated with each other, either heavily or lightly, while retaining the variation present in the data set up to the maximum extent.

Linear discriminant analysis (LDA) is another dimensionality reduction method that is used to separate two groups/classes. The main idea of LDA is to maximize the separability between the two groups to enable a better classification. LDA is similar to PCA, but it focuses on maximizing the separability among known categories by creating a new linear axis and projecting the data points on that axis [3].

Generalized discriminant analysis (GDA) is utilized for multiclass classification problems. Due to the large variations in the patterns of various classes, there is usually a considerable overlap between some of these classes in the feature space. In this situation, a feature transformation mechanism is used that can minimize the between-class scatter [4].

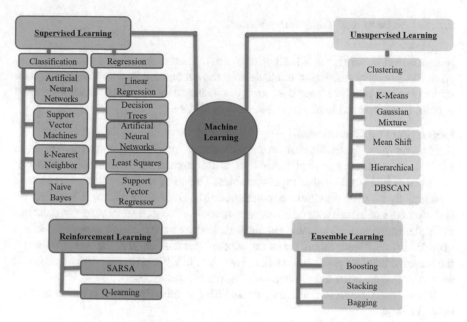

Fig. 6.5 Overview of ML models and algorithms

Learning modeling: Once the preprocessing stage is completed, the data is processed in the (machine) learning modeling stage. A detailed overview of the various ML models is presented in the following section. ML algorithms are divided into four subgroups, as shown in Fig. 6.5:

- Supervised learning
- Unsupervised learning
- Reinforcement learning
- Ensemble learning

Postprocessing: Postprocessing enables the data scientist to determine how effective data provided to the ML algorithm were in modeling the process under study. Combination, model output statistics (MOS), and error correction methods are the most commonly used postprocessing techniques. In some cases, it is essential to combine various ML models to improve predictions. MOS determines the systematic dependencies and improves the overall performance of the learning model. Error correction techniques may be optionally be used for reducing the systematic errors in the modeling.

Performance analysis: Finally, performance analysis determines how effective the model is in predicting a trial data set by comparing the observed and predicted values. Mean absolute percentage error (MAPE), root mean square error (RMSE), and correlation analysis are the most used performance metrics used for the ML algorithms and models.

6.2.1 Supervised Learning Models

In supervised learning, the ML algorithms are trained using labeled data. Each data point has a label or collection of labels, and the supervised learning algorithm constructs a model by classifying the data into sets based on the labels. Once the model is built, it can be used to classify new, unlabeled data.

Regression and Classification

Regression and classification are two major prediction methods used in ML. Regression is commonly used to build numerical relationships between observed outputs and a set of input variables. The complexity level of regression is significantly lower in comparison to advanced ML and AI algorithms. Classification is the process of identifying or detecting a model to assist in separating the data into multiple categorical classes and subclusters. For the classification approach, data is grouped into different labels based on selected parameters, then the output, that is, the selected label, is predicted for the given data [5]. Visualization techniques such as cluster graphs often help in determining the classes and subclusters.

Widely used supervised learning models for classification and regression are discussed below.

Linear regression: Linear regression (LR) is the simplest form of regression method. LR aims to estimate the relationship between two variables by fitting a linear equation to the observed data set [6]. The first variable is the dependent and target variable, and the second is the independent or explanatory variable. Simple LR models accommodate only one explanatory variable, but multiple LR models may deal with more than one explanatory variable. The target variable is predicted by using only one predictor. In that case, the mathematical model for describing the LR is given by

$$Y = a + bX \qquad (6.1)$$

where X represents the explanatory variable, and Y represents the target variable. Parameters a and b are the weights of the model. b is the slope of the line, and a is the Y-intercept.

Decision tree learning: Decision tree learning is a supervised learning algorithm used for classification and regression problems or for creating decision support tools based on tree-like decision models. A decision tree relies on a set of conditional control statements, for example, if-else conditions that allocate instances and decisions from the parent (decision node) to the child node (leaf node). Regression decision trees yield numeric responses, while classification decision trees' responses are in Boolean form, that is, either true or false. The first node in a decision tree is usually named the Root Node. Figure 6.6 demonstrates a sample decision tree [7].

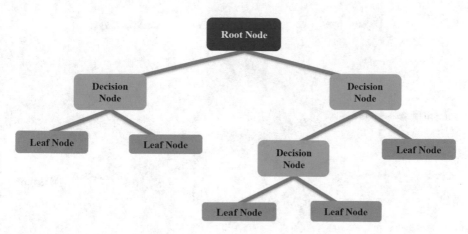

Fig. 6.6 Explanation of decision trees

Artificial neural networks: Artificial neural network (ANN) algorithms are inspired by biological neural systems assuming that the best way to mimic human inference and decision-making is through structurally similar systems built-in software. It is used for classification as well as regression. ANNs consist of multiple artificial neurons named perceptrons and layers. A simplified version of ANNs can consist of input, hidden, and output layers. Each connection or synapse between artificial neurons is responsible for transmitting signals in the form of real numbers from the preceding layer as output to the following layer as input. A weighted sum of inputs computes the output of each layer. All weights in an ANN and the associated threshold are set to random values during training. The adjustment process of the thresholds and weights continues during the training until the same data labels result in similar outcomes [8]. If the transmitted number is above the threshold value or activation function, the corresponding neuron opens the gate and allows the number to pass through to the next neuron. If the number is below the threshold value, the neuron will not pass the output to the next neuron's input.

Activation functions are used to determine the output of the neuron. An activation function is constructed so that the output of each neuron is normalized to be between 0 and 1 or −1 and 1. Activation functions can be in the form of a binary step, linear, sigmoid, or tanh function. The propagation process is repeated until the signal is transmitted via various hidden layers to the output layer (Fig. 6.7).

ANNs can be constructed having the following architectures:

- Feedforward NN (FNN)
- Multilayer perceptron (MLP)
- Convolutional neural networks (CNN)
- Long short-term memory (LSTM)
- Recurrent NN (RNN)

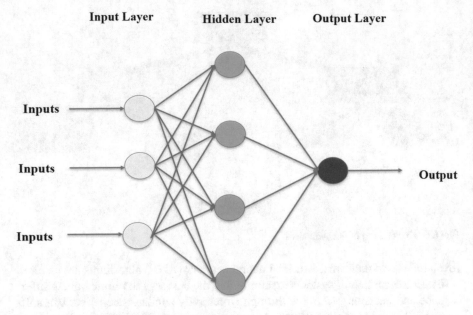

Input Layer   **Hidden Layer**   **Output Layer**

Inputs

Inputs   Output

Inputs

Fig. 6.7 A simplified demonstration of an NN

Least squares: The least-squares method is a form of regression analysis that selects the best parameter estimate by minimizing the sum of the squared errors between the predicted and observed target variable, thereby identifying the best-fit line. Each data point characterizes the relationship between an unknown target variable and a known explanatory variable. But the data points may not have equal importance in a weighted least square problem [8, 9].

Support vector machines: Support vector machines (SVMs) are supervised learning models that are used for classification or regression analysis. They use the data to represent a hyperplane as a decision surface nearest to the data set. SVMs are categorized based on their kernel method types. A kernel method provides a systematic approach to training the ML algorithm. The selection of the kernel function type is a key feature for SVM algorithms in addition to the data input selection. The training function automatically determines the kernel function.

In SVM, training data does not necessarily have to be divided into a validation and a training data set. The training process for SVM has two main steps:

- Mapping the input data (predictors) to a high-dimensional feature space where the kernel trick function is evaluated. The trick function is a pairwise similarity function that helps to save time and storage by finding an appropriate kernel function.
- Solving a quadratic optimization problem to find the optimal parameters and computing the decision boundaries. SVMs are designed to maximize the margin between clustered labeled training data points separated by support vectors. The

quadratic optimization problem finds the optimal decision boundaries based on the formed hyperplanes around the labeled data points.

Support vector regressor: Support vector regressor (SVR) is a subcategory of SVMs. SVR is used for classification and at the same time for regression analysis with some adjustment. The basic structure of SVM can be slightly modified to deal with the regression problem using SVR that uses all the features of SVM, that is, attempts to determine a curve based on given data points. However, SVR aims to find a match between related vectors and the position on the curve instead of a decision boundary as in SVM. Finding the best match between the data points and the actual points is accomplished by support vectors. The process continues until the distance between the support vectors and the regressed curve is maximized. During SVR, the vectors that don't offer support are discarded as to statistical outliers.

k-Nearest neighbors: The k-nearest neighbors (k-NN) technique is a nonparametric method widely used for classification and regression problems in the AI and ML domains. k is a positive integer and typically small, that is, 1–10. The k-NN model input consists of the k closest training examples in the feature space. For the classification approach, the model output presents a categorical membership. A majority vote of its own neighbors decides the membership of an object. The object is assigned to the categorical membership with its k-nearest neighbors. For the regression approach, the model output is the average of its k-nearest neighbors' values for the object [10].

Naïve Bayes: Bayes' theorem is applied when computing the probability that a newly observed data set belongs to a given class and given existing data set. Naïve Bayes is a classification algorithm based on Bayes' theorem for both binary and multiclass classification problems. The name naïve is used within the approach because the algorithm assumes the features that are injected into the model are independent of each other. The difficulty levels of the computations are simplified to increase the traceability of the algorithm.

6.2.2   Unsupervised Learning

Unsupervised learning is a subcategory of ML that considers the previously unde-tected patterns in a specific data set without any labels. It mainly deals with unla-beled data. Unsupervised learning algorithms primarily aim to discover the similarities in data by clustering similar data points together, thereby discovering patterns in the data set. The following algorithms are commonly used in unsuper-vised ML for clustering:

- k-means
- Gaussian mixture
- Mean shift clustering

- Hierarchical clustering
- DBSCAN

k-Means: The *k*-means algorithm is the simplest unsupervised learning algorithm, aiming to minimize the sum of distances between instances in a specific cluster and their corresponding centroids, that is, averages. The data set in a large cluster is broken into a predetermined number (*k*) of smaller data clusters by the *k*-means algorithm. First, a centroid presenting a data point at the center of each cluster is defined, that is, considering Euclidean distance for each cluster. Then, each data point in the cluster is assigned to its nearest centroid based on the squared Euclidean distance [11].

Gaussian mixture: The Gaussian mixture algorithm is like *k*-means in terms of a clustering method of unsupervised data. However, it offers some advantages over *k*-means. The Gaussian mixture model clusters the data inputs as using a mixture of multiple Gaussian functions, not just by their mean as in k-means [12]. It provides a flexible class of probability distributions and explains variation with the cluster data. First, the model starts with parameters describing each cluster, such as mean, variance, and size, and calculates the probability distribution. Second, the model selects a mixture component with a probability and samples from the selected component's Gaussian.

Mean shift clustering: Mean shift clustering is a sliding-window- and a centroid-based algorithm that intends to find dense areas of data points in a large data set. The key function of the algorithm is to assign each data point to the closest cluster centroid, and this process continues iteratively until finding where the most points are located for each data point, that is, the cluster center. Once the algorithm decides that the data point is placed where the most points are, it is assigned to a cluster or corresponding group [13].

Hierarchical clustering: The hierarchical clustering algorithm is a form of cluster analysis that attempts to build a hierarchy of cluster stages. There exist two types of hierarchical clustering algorithms:

- Agglomerative (bottom-up approach)
- Divisive (top-down approach)

The agglomerative approach starts with assuming that each point is a cluster and then repeatedly combining the two "nearest" clusters into one, while the divisive approach starts with one cluster and recursively split it. However, both approaches are pretty similar in terms of operation. The algorithm starts with all the data points assigned to a cluster by its centroid, that is, the average of its points. Then two nearest clusters are combined into the same cluster. The process continues until there is only a single cluster left or the next merge would create a cluster with low cohesion [14].

DBSCAN: DBSCAN is a density-based clustering algorithm and stands for density-based spatial clustering of applications with noise. It is widely used primarily for finding outliers in a larger data set. The algorithm discovers randomly shaped clusters based on the density of data points in different regions. Density is the key measure for the algorithm to separate regions, that is, locating that outliers between the high-density clusters. DBSCAN uses two parameters to determine clusters, that is, MinPts (the minimum number of data points) and Eps (the distance). MinPts is required for an area to be considered high density. Eps is used to determine whether or not a data point is in the same area as other data points [15].

6.2.3 Reinforcement Learning

Reinforcement learning (RL) enables agents to learn from their own actions using trial and error by interacting with an environment that provides rewards. Figure 6.8 provides an overview of the reinforcement learning process. Agents are intelligent programs and the decision-making element in an RL system. Environments are the surrounding domain where agents' actions are performed. Actions are the tasks performed by the agents in the environment. Agents are trained using rewards [16]. There are two well-known reinforcement learning algorithms, that is, Q-learning and state-action-reward-state-action (SARSA).

Q-learning: Q-learning is a basic form of RL and uses action values (Q-values) to improve the behavior of the learning agent by telling the agent what action shall be taken under what conditions. The "Q" in Q-learning stands for "quality." Q-learning uses a trial-and-error-based approach and aims to learn a policy that maximizes the rewards. It consists of several characteristics, such as an input and output system, reward, an environment, a Markov decision process, training and interference, possible states (S), and actions (A). The algorithm continuously updates its policy as it learns more and more about is the environment and updates the value function based on an equation, the Bellman optimality equation [17].

SARSA: SARSA is an on-policy and model-free algorithm, which is a variation of the Q-learning algorithm based on Markov Decision Process (MDP) policy. SARSA stands the tuple of (S, A, R, S1, A1), that is, the current state (S), action

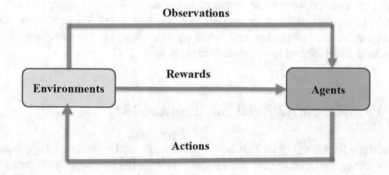

Fig. 6.8 Reinforcement learning overview

(A), reward (R), next state (S1), and action taken (A1). The difference between SARSA and Q-learning that the Q-value is updated, taking into account the action, A1 performed in the state, S1 in SARSA as opposed to Q-learning where the action with the highest Q-value in the next state, S1 is used to update Q-table [18].

6.2.4 Ensemble Learning

Ensemble learning algorithms combine multiple learning algorithms to provide better predictive performance or to generate an optimal predictive model. Boosting, bagging, and stacking are the subcategories of ensemble learning algorithms.

Boosting algorithms: Boosting algorithms are an ensemble model widely used algorithm in data science competitions. They are designed to reduce the bias and variance and often offer better performance than simpler models, such as decision trees (DT) and logistic regression (LR). The goals of boosting algorithms are to decrease the bias error and build a strong classifier from a number of weaker classifiers. They improve the prediction outputs by training a sequence of weak models, each compensating for the weaknesses of its predecessors [19].

Bagging algorithms: Bagging algorithms, which stands for bootstrap aggregating, are a parallel ensemble model. They attempt to improve the accuracy and the performance of the ML algorithms for solving regression and classification problems. Bagging algorithms reduce variance to avoid overfitting, which is a common modeling error in statistics and ML. In a bagging algorithm, additional data is generated by random sampling with replacement from the original data set in each training stage. Therefore, the size of the training data set is increased while the variance decreases, and the prediction is narrowly tuned to an expected outcome [20].

Stacking algorithms: Stacking algorithms combine the predictions from multiple ML models on the same data set. However, they mainly differ from bagging and boosting on two points. First, in terms of the type of weak learners, stacking algorithms focus on heterogeneous weak learners while bagging and boosting focus on homogeneous weak learners. Second, in terms of the type of learning, stacking algorithms learn in parallel and combine weak learners by training a meta-model while bagging and boosting run weak learners sequentially and combine them following deterministic algorithms [21].

6.3 Explainable Artificial Intelligence (XAI)

Artificial intelligence (AI) models have been widely used in smart grid applications, such as energy generation [22] and load forecasting [23], energy management and control [24], and predictive maintenance [25]. However, there are still many

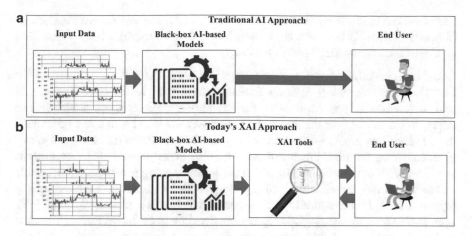

Fig. 6.9 (**a**) Traditional AI approach and (**b**) today's XAI approach

concerns with using AI, which is considered a black box due to lack of model explainability, transparency, and trustworthiness. Decision-makers and domain experts must be able to understand and trust the results obtained from AI models, and they cannot trust them unless the process by which the results were obtained is understandable in terms of the problem domain. To address these concerns, explainable artificial intelligence (XAI) algorithms have been developed to open the AI black box and to help decision-makers understand the model results. XAI can additionally foster the development of more accurate and sophisticated AI models by increasing model explainability and transparency [26]. Existing XAI approaches utilize post hoc methods to understand a previously trained model and to explain its predictions. Figure 6.9 compares the traditional AI approach with the XAI approach. The traditional AI approach provides outputs through trained models without any feedback to human users. On the other hand, the XAI approach provides not only predictions but also an explanation of why the predictions turned out to be a certain way. For example, XAI can answer questions such as why the algorithm did that, can the results be trusted, is it possible to correct the results, etc. Understanding through transparency is crucial for automated decision-making and for regulatory compliance. AI model results can be understood by decision-makers and domain experts through the information and additional results provided by XAI approaches, in contrast to the black box of traditional machine learning algorithms. XAI also allows interaction between domain experts and the algorithm to be able to learn and correct the errors as well as to re-enact the calculations on-demand.

XAI methods are an emerging and expanding research topic in the machine learning field due to the need to interpret, explain, debug, and test black box AI models. Several XAI tools were developed by AI community and companies. They include LIME (Local Interpretable Model-Agnostic Explanations) [27], DeepLIFT [28], Skater [29], SHAP (Shapley Additive exPlanation) [30], ELI5 [31], AIX360 [32], MLxtend (machine learning extensions) [33], InterpretML [34], Rulex

Explainable AI [35], TreeInterpreter [36], Alibi [37], CEM (Contrastive Explanation Method) [38], H2O [39], and XAI – eXplainableAI Framework [40]. These XAI algorithms are summarized in this section.

Local Interpretable Model-Agnostic Explanation (LIME)
Local Interpretable Model-Agnostic Explanations (LIME) is widely used as an XAI tool. It provides a local explanation for any classifier or regressor model and an interpretable model over the interpretable representation to be able to explain how the AI black box model works locally [41]. LIME focuses on training local surrogate models on perturbed inputs instead of training them globally [42]. Local means how the model behaves in the vicinity of the instance being predicted. Interpretable means explaining the classifier and regressor model in terms of the interpretable representation. Model-agnostic explanation means it can explain any model [43], such as linear regression, logistic regression, decision trees, Naive Bayes, k-nearest neighbors, and others. LIME can be applied to textual, image, and tabular data [44]. The core LIME XAI tool was written in Python and is available from a GitHub repository [45].

DeepLIFT (Deep Learning Important FeaTures)
Deep Learning Important FeaTures (DeepLIFT) is an XAI tool that uses a back-propagating explanation model to be able to explain the prediction with the important part of each input. DeepLIFT is a different-from-reference method consisting of the following steps: (1) input and output reference values are defined, (2) the difference between an output value and its reference value is calculated, (3) the difference is attributed to the various input-reference differences, and (4) the prediction output is calculated as the sum of all contributions. The DeepLIFT tool provides an importance score to individual inputs based on their positive and negative contributions as well as bias terms. Results obtained with DeepLIFT can be plotted as a saliency map. It is an efficient tool in terms of computation and power consumption since the importance scores can be obtained in a single backward pass after making a prediction [46].

Skater
Skater is an open-source unified framework developed in Python for model interpretation. Skater provides both global (i.e., inference based on a complete data set) and local (i.e., inference about an individual prediction) explanations for an AI black box. This framework can be used for various models, including natural language processing (NLP), ensemble, and image recognition models. It supports two types of model interpretation, post hoc interpretation and natively interpretable models. For the post hoc interpretation model, a black box model is trained to solve a supervised learning problem, and the explanation model provides a visual or textual representation showing the relationship between input data and the predicted result(s). This helps a decision-maker or domain expert understand the model by answering some questions, for example, why a certain output is favored over another. For the natively interpretable model, the predictive model having an explanator function does not need an additional explanation tool since it is designed

as a transparent model being interpretable [47]. Skater is a Python package available at a GitHub repository with many examples [48].

SHapley Additive exPlanation (SHAP)

SHapley Additive exPlanation (SHAP) uses Shapely values [49] to determine the impact of each feature on the prediction, either positively or negatively. SHAP was introduced in the study [50] to provide an explanation for individual predictions through Shapley values, that is, feature impact scores. This method computes Shapley values from coalitional game theory, which represent the average marginal contribution of a feature value over all possible coalitions, that is, all possible predictions, for instance, using all possible combinations of inputs. It can also provide (1) local interpretability by providing an explanation for a certain instance and (2) global interpretability by averaging local explanations. In addition, SHAP values can be calculated for any tree-based model [51]. SHAP is widely used for regression and classification models with tabular data to explain which features contributed positively or negatively to the prediction. The SHAP package is mostly written in Python and offers various explainers, such as TreeExplainer, DeepExplainer, GradientExplainer, LinearExplainer, and KernelExplainer.

Explain Like I'm 5 (ELI5)

Explain Like I'm 5 (ELI5) was developed to be able to debug machine learning classifiers, verify the accuracy of machine learning models, and explain their predictions with useful visualization tools using Scikit-learn and XGBoost. ELI5 provides an API for many frameworks and packages; however, it supports a limited set of models, primarily tree-based and other parametric and linear models. The prediction is described as the sum of each feature impact and the bias term [52]. ELI5 provides explanations for all the weights and predictions of Scikit-learn linear classifiers and regressors and decision trees and shows the importance of each feature. In addition, ELI5 supports text processing utilities from Scikit-learn and can highlight text data accordingly [53].

The AI Explainability 360 (AIX360)

The AI Explainability 360 (AIX360) is an open-source Python XAI tool developed by IBM that provides interpretation and explanation for data sets and machine learning models. It supports a variety of explanatory algorithms, including data, local, and direct global explanations. These explanatory algorithms include data explanation, local/global post hoc explanation, and local/global direct explanation. It also supports two explanatory metrics, namely faithfulness and monotonicity [54]. The faithful metric evaluates the correlation between the importance assigned to features by the interpretability algorithm and the importance of each feature for the performance of the predictive model, while the monotonicity metric measures the importance of each feature on the model performance. AIX360 provides a unified, flexible, and easy-to-use programmatic interface supporting a variety of explanatory techniques required by both data scientists and algorithm developers.

Machine Learning Extensions (MLxtend)

Machine Learning Extensions (MLxtend) was primarily developed as an extension of Sci-kit learning library in Python to provide researchers and data scientists with easy access to commonly used machine learning libraries. It consists of a variety of functions, including sequential feature selection algorithms (forward, backward, forward floating, and backward floating selection) as well as visualization functions allowing users to inspect the estimated predictive performance, including performance intervals for different feature subsets. It supports a completely automated feature extraction and selection tool, as well as model evaluation techniques to allow users to compare the performance of different machine learning algorithms to each other [55]. MLxtend has a high potential to be extended its capabilities for XAI community due to being developed in Python.

InterpretML

InterpretML developed by Microsoft is an open-source Python-based framework for training interpretable machine learning models and explaining black box systems. InterpretML supports both interpretability models, that is, glass box and black box. Glass box covers machine learning models like linear models, rule lists, and generalized additive models, already designed for interpretability, and other techniques for explaining black box models such as Partial Dependence and LIME. InterpretML also includes a new glass box model interpretability approach, Explainable Boosting Machine (EBM), which provides an accuracy table containing the contribution of each feature to the prediction. It also allows users to compare interpretability algorithms through a unified API and a built-in visualization tool. InterpretML has the following key advantages: (1) easy to compare multiple algorithms, (2) uses reference algorithms and visualizations, (3) leverages the open-source ecosystem, and (4) component extension [56]. InterpretML is developed in Python and available at the GitHub repository [57].

TreeInterpreter

TreeInterpreter is a Python package for interpreting decision trees and random forest predictions in the Sci-kit learning library. The predictions are easily interpretable through TreeInterpreter. Each prediction is calculated with a sum of contributions from each feature, that is, *prediction = Bias + Sum of each features' contributions*. The bias term represents the training set mean, while the other contributions represent the impact of each feature individually. The package allows users to sort the feature contributions by their absolute impact to see the contributions of the most and least important feature. This also helps in understanding why a prediction is high or low, for example, why the predicted value is much higher since some features may actually have a very large positive impact, or why the predicted value is much lower since some features may actually have a very large negative impact. Random forest predictions can be interpreted straightforwardly, leading to a similar level of interpretability as linear models.

Alibi

Alibi is an open-source Python library developed by Seldon, which provides the explanation for predictions of machine learning models. Alibi focuses on inspection and interpretation of classification and regression models and offers multiple algorithms for both global and local interpretability. It consists of an explanatory library with many algorithms and an explanation API allowing interaction with other tools. The explanation API takes inspiration from Scikit-learn and supports distinct initialize, fit and explain steps. These algorithms provide instance-specific scores measuring the model confidence for making a particular prediction. Alibi's new version is integrated to Kernel SHAP, which is a black box model interpreter widely used for regression and classification problems [58].

Contrastive Explanation Method (CEM)

Contrastive Explanation Method (CEM) was developed by IBM and is widely used to compute contrastive explanations justifying the classification of input for image and tabular data. CEM can generate black box model explanations in terms of pertinent positives (PP) and pertinent negatives (PN). For PP, it finds what should be minimally and sufficiently present to justify its classification. On the other hand, PN identifies what should be minimally and necessarily absent from the explained instance to maintain the original prediction. An example of the area of image recognition is input is considered minimally and sufficiently present if it is an important object pixel in an image necessary to justify its classification, while the pixel is considered minimally and necessarily absent if it is a certain background pixel [59].

H2O Driverless AI

H2O Driverless AI developed by H2O.ai is an open-source machine learning platform to explain modeling results through K-LIME, Shapley, Variable Importance, Decision Tree, Partial Dependence with the MLI (Machine Learning Interpretability) capability in a cloud environment. H2O.ai has improved this platform by providing XAI functionalities with additional features [60]. H2O Driverless AI is an industry-leading machine learning platform that empowers researchers to be more productive by accelerating workflows with automatic feature engineering, customizable user-defined modeling methods, and automatic model deployment, among many other leading-edge capabilities. The platform allows data scientists to utilize different techniques and methodologies for interpreting and explaining the results of its models, as well as includes interfaces for R, Python, Scala, Java, JSON, and CoffeeScript/JavaScript [61].

XAI – eXplainableAI Framework

XAI – eXplainableAI framework developed by The Institute for Ethical AI & ML is an open-source machine learning framework that focuses on promoting AI explanation and transparency. The alpha version is designed to analyze and evaluate data and AI models. The framework provides three main steps: (1) data analysis, (2) model evaluation, and (3) production monitoring. XAI implements a glass box model that has many visualization functions. The framework is developed in Python and available at GitHub repository [62].

Table 6.2 The summary of XAI tools with their features

XAI tool	Intrinsic/post hoc (I/H)	Global/local (G/L)	Model agnostic/ specific (A/S)	Programming language
LIME	H	L	A	Python/R
DeepLIFT	H	G/L	S	Python
Skater	I	G/L	A	Python
SHAP	H	G^a/L	A	Python/R
ELI5	H	G/L	S	Python
AIX360	I	G/L	A	Python
MLxtend	H	G	S	Python
InterpretML	I	G/L	A	Python/R
TreeInterpreter	I	G/L	S	Python
Alibi	H	G/L	A	Python
CEM	H	L	A	Python
H2O	I	G/L	A	Python
XAI	H	G	A	Python

aGlobal interpretability by averaging local explanations

The widely used XAI tools with their features are summarized in Table 6.2 in terms of interpretability (intrinsic vs. post hoc), the scope of interpretability (global vs. local), model interpretability category (model agnostic vs. model-specific), and development in programming. Intrinsic interpretability refers to the model being interpretable itself, also called transparency, and answers the question of how the model works. Post hoc interpretability refers to the model explanation after model training, which answers the question of what else the model tells us [63]. Global-model interoperability provides a complete view on the inner working of the model, the data features, and all learned components, for example, weights, parameters, etc., as well as showing the machine-learned relationships between the predictions and the input variables across entire data, but global interoperability can be highly approximate in some cases. Local-model interoperability provides detailed information about the relationship between a selected prediction and the input variables and tries to understand how the model made a decision at its prediction. The model-agnostic interpretation refers to the method that can be applied to different types of machine learning algorithms. A model-specific interpretation refers to methods that can be applied to only a single type or class of AI algorithm.

6.4 Summary

This chapter provides a brief and compact overview of the basic terminologies and definitions in big data (BD), machine learning (ML), and artificial intelligence (AI). BD focuses on collecting, cleansing, storing, analyzing, and extracting information from, align with interpreting large data sets in science and businesses from healthcare, media, and energy systems to defense. AI and ML-based algorithms are widely used in the research to perform tasks that typically require human intelligence, such as driving cars, translating speech, and image recognition. Machine learning algorithms are generally split into four main categories: supervised learning, unsupervised learning, reinforcement learning, and ensemble learning. Supervised learning-based ML algorithms are trained using labeled data, such as linear regression, decision tree learning, artificial neural networks, least-squares, support vector machines, support vector regressor, k-nearest neighbors, and Naïve Bayes. Unsupervised learning-based algorithms mainly deal with unlabeled data to discover the similarities in data by clustering similar data points together, which includes k-means, Gaussian mixture, mean shift clustering, hierarchical clustering, DBSCAN, and others. Reinforcement learning (RL) is another type of machine learning algorithm by interacting with an environment using trial and error with a numerical reward. There are two well-known reinforcement learning algorithms, namely, Q-learning and state-action-reward-state-action (SARSA). Ensemble learning is a machine learning-based algorithm combining multiple learning algorithms to provide better predictive performance or to generate an optimal predictive model. Ensemble learning algorithms can be classified into boosting, bagging, and stacking algorithms.

The traditional AI function as a black box model typically does not provide decision-makers and domain experts with any guidance as to why a particular decision was made. This lack may often lead decision-makers and domain experts to question the results and ultimately reject them because they cannot explain the approach to stakeholders, politicians, and others who want to know why a particular decision was made. Explainable AI (XAI) provides a more transparent and explanatory approach, thereby rendering the decision more acceptable to people who need to explain why a particular decision was made to others. In addition, this chapter provides a comprehensive overview of currently available Explainable AI (XAI) tools and summarizes them in terms of interpretability (intrinsic vs. post hoc), the scope of interpretability (global vs. local), model interpretability category (model agnostic vs. model-specific), and development in programming.

References

1. Sagiroglu, S. and Sinanc, D., 2013, May. Big data: A review. In 2013 international conference on collaboration technologies and systems (CTS) (pp. 42–47). IEEE.
2. Gandomi, A. and Haider, M., 2015. Beyond the hype: Big data concepts, methods, and analytics. International journal of information management, 35(2), pp. 137–144.
3. PCA vs LDA vs T-SNE, https://medium.com/analytics-vidhya/pca-vs-lda-vs-t-sne-lets-understand-the-difference-between-them-22fa6b9be9d0, (Accessed Aug. 29, 2021).
4. Singh, S. and Silakari, S., 2009. Generalized discriminant analysis algorithm for feature reduction in cyber attack detection system. arXiv preprint arXiv:0911.0787.
5. ML-Classification vs Regression, https://www.geeksforgeeks.org/ml-classification-vs-regression/, (Accessed Aug. 29, 2021).
6. Montgomery, D.C., Peck, E.A. and Vining, G.G., 2012. Introduction to linear regression analysis (Vol. 821). John Wiley & Sons.
7. Hall, L.O., Chawla, N. and Bowyer, K.W., 1998, October. Decision tree learning on very large data sets. In SMC'98 Conference Proceedings. 1998 IEEE International Conference on Systems, Man, and Cybernetics (Cat. No. 98CH36218) (Vol. 3, pp. 2579–2584). IEEE.

8. Friedman, J., Hastie, T. and Tibshirani, R., 2001. The elements of statistical learning (Vol. 1, No. 10). New York: Springer series in statistics.
9. Cali, Umit, and Claudio Lima. "Energy informatics using the distributed ledger technology and advanced data analytics." Cases on Green Energy and Sustainable Development. IGI Global, 2020. 438–481. (2)
10. Rahman, I., Kuzlu, M. and Rahman, S., 2018. Power disaggregation of combined HVAC loads using supervised machine learning algorithms. Energy and Buildings, 172, pp. 57–66.
11. https://analyticsindiamag.com/most-popular-clustering-algorithms-used-in-machine-learning/, (Accessed Aug. 29, 2021).
12. https://towardsdatascience.com/gaussian-mixture-models-d13a5e915c8e, (Accessed Aug. 29, 2021).
13. https://www.geeksforgeeks.org/ml-mean-shift-clustering/, (Accessed Aug. 29, 2021).
14. https://www.analyticsvidhya.com/blog/2016/11/an-introduction-to-clustering-and-different-methods-of-clustering/, (Accessed Aug. 29, 2021).
15. https://www.freecodecamp.org/news/8-clustering-algorithms-in-machine-learning-that-all-data-scientists-should-know/, (Accessed Aug. 29, 2021).
16. Sutton, R.S. and Barto, A.G., 2018. Reinforcement learning: An introduction. MIT press.
17. https://towardsdatascience.com/a-beginners-guide-to-q-learning-c3e2a30a653c, (Accessed Aug. 29, 2021).
18. https://medium.com/swlh/introduction-to-reinforcement-learning-coding-sarsa-part-4-2d64d6e37617, (Accessed Aug. 29, 2021).
19. https://towardsdatascience.com/boosting-algorithms-explained-d38f56ef3f30, (Accessed Aug. 29, 2021).
20. https://www.mygreatlearning.com/blog/bagging-boosting, (Accessed Aug. 29, 2021).
21. https://towardsdatascience.com/ensemble-methods-bagging-boosting-and-stacking-c9214a10a205, (Accessed Aug. 29, 2021).
22. Mellit A, Kalogirou SA. Artificial intelligence techniques for photovoltaic applications: A review. Progress in energy and combustion science. 2008 Oct 1;34(5):574–632.
23. Raza MQ, Khosravi A. A review on artificial intelligence based load demand forecasting techniques for smart grid and buildings. Renewable and Sustainable Energy Reviews. 2015 Oct 1;50:1352–72.
24. Zhou H, Rao M, Chuang KT. Artificial intelligence approach to energy management and control in the HVAC process: an evaluation, development and discussion. Developments in Chemical Engineering and Mineral Processing. 1993;1(1):42–51.
25. De Benedetti M, Leonardi F, Messina F, Santoro C, Vasilakos A. Anomaly detection and predictive maintenance for photovoltaic systems. Neurocomputing. 2018 Oct 8;310:59–68.
26. Utility Dive, How does AI improve grid performance? No one fully understands and that's limiting its use, [Online]. Available: https://www.utilitydive.com/news/how-does-ai-improve-grid-performance-no-one-fully-understands-and-thats-l/566997/
27. Ribeiro MT, Singh S, Guestrin C. "Why should i trust you?" Explaining the predictions of any classifier. In Proceedings of the 22nd ACM SIGKDD international conference on knowledge discovery and data mining 2016 Aug 13 (pp. 1135–1144).
28. Shrikumar, A., Greenside, P. and Kundaje, A., 2017. Learning important features through propagating activation differences. arXiv preprint arXiv:1704.02685.
29. Skater: Python library for model interpretation/explanations. GitHub repository, https://github.com/oracle/Skater
30. Siddhartha M, Maity P, Nath R. Explanatory Artificial Intelligence (Xai) In The Prediction Of Post-Operative Life Expectancy In Lung Cancer Patients. International Journal of Scientific Research. 2020 Jan 24;8(12).
31. "ELI5." https://github.com/TeamHG-Memex/eli5, 2019. [Online; Accessed 2019-06-04].
32. Arya, V., Bellamy, R.K., Chen, P.Y., Dhurandhar, A., Hind, M., Hoffman, S.C., Houde, S., Liao, Q.V., Luss, R., Mojsilovic, A. and Mourad, S., 2020. Ai explainability 360: An extensible toolkit for understanding data and machine learning models. Journal of Machine Learning Research, 21(130), pp. 1–6.

33. Raschka S. MLxtend: Providing machine learning and data science utilities and extensions to Python's scientific computing stack. Journal of open source software. 2018 Apr 22;3(24):638.
34. InterpretML, Explain Your Model with Microsoft's InterpretML, https://medium.com/analytics-vidhya/explain-your-model-with-microsofts-interpretml-5daab1d693b4
35. Rulex Explainable AI (XAI), https://www.rulex.ai/rulex-explainable-ai-xai/
36. Félix Revert, Interpreting Random Forest and other black box models like XGBoost, https://towardsdatascience.com/interpreting-random-forest-and-other-black-box-models-like-xgboost-80f9cc4a3c38
37. Seldon, Alibi, https://docs.seldon.io/projects/alibi/en/latest/overview/getting_started.html.
38. IBM, Contrastive Explanation Method (CEM), https://github.com/IBM/Contrastive-Explanation-Method
39. Machine Learning Interpretability (MLI), https://github.com/h2oai/mli-resources
40. XAI – The eXplainable AI Framework, https://ethical.institute/xai.html
41. Joshua Poduska, SHAP and LIME Python Libraries: Part 1 – Great Explainers, with Pros and Cons to Both, https://blog.dominodatalab.com/shap-lime-python-libraries-part-1-great-explainers-pros-cons/
42. Sumit Saha, local interpretable model-agnostic explanations (lime) – the eli5 way, https://medium.com/intel-student-ambassadors/local-interpretable-model-agnostic-explanations-lime-the-eli5-way-b4fd61363a5e
43. Marco Tulio Ribeiro, LIME – Local Interpretable Model-Agnostic Explanations, https://homes.cs.washington.edu/~marcotcr/blog/lime/
44. Sumit Saha, local interpretable model-agnostic explanations (lime) – the eli5 way, https://medium.com/intel-student-ambassadors/local-interpretable-model-agnostic-explanations-lime-the-eli5-way-b4fd61363a5e
45. https://github.com/marcotcr/lime
46. Eriksson, T., 2020. Occlusion method to obtain saliency maps for CNN.
47. Skater Overview, https://oracle.github.io/Skater/overview.html
48. Skater, https://github.com/oracle/Skater
49. Hart, S., 1989. Shapley value. In Game Theory (pp. 210–216). Palgrave Macmillan, London.
50. Lundberg, Scott M., and Su-In Lee. "A unified approach to interpreting model predictions." Advances in Neural Information Processing Systems. 2017
51. towardsdatascience, Explain Your Model with the SHAP Values, https://towardsdatascience.com/explain-your-model-with-the-shap-values-bc36aac4de3d
52. On Click 260, nterpretable Machine Learning With Lime+ELI5+SHAP+InterpretML, https://www.onclick360.com/interpretable-machine-learning-with-lime-eli5-shap-interpret-ml/
53. ELI5, ELI5, https://github.com/TeamHG-Memex/eli5
54. AI Explainability 360, https://aix360.readthedocs.io/en/latest/
55. Raschka, S., 2018. MLxtend: providing machine learning and data science utilities and extensions to Python's scientific computing stack. Journal of open source software, 3(24), p.638.
56. Nori, H., Jenkins, S., Koch, P. and Caruana, R., 1909. InterpretML: A Unified Framework for Machine Learning Interpretability (2019). arXiv preprint arXiv:1909.09223.
57. InterpretML - Alpha Release, https://github.com/interpretml/interpret
58. Alibi Explain, https://github.com/SeldonIO/alibi
59. Dhurandhar, A., Chen, P.Y., Luss, R., Tu, C.C., Ting, P., Shanmugam, K. and Das, P., 2018. Explanations based on the missing: Towards contrastive explanations with pertinent negatives. In Advances in Neural Information Processing Systems (pp. 592–603).
60. Explainable AI: The Next Best Thing in Digital Health, https://ekare.ai/explainable-ai-the-next-best-thing-in-digital-health/
61. An Introduction to eXplainable AI with H2O Driverless AI, https://andisama.medium.com/an-introduction-to-explainable-ai-with-h2o-driverless-ai-2a9e8f27e03f
62. XAI – An eXplainability toolbox for machine learning, https://github.com/EthicalML/xai
63. Carvalho, D.V., Pereira, E.M. and Cardoso, J.S., 2019. Machine learning interpretability: A survey on methods and metrics. Electronics, 8(8), p. 832.

Chapter 7
Applications of Artificial Intelligence in the Energy Domain

While the electric power grid of the twentieth century was fairly simple, with one-directional power flow from centralized generating plants through the transmission grid and the distribution grid to end user loads, the power grid of the twenty-first century has become increasingly complex. The incorporation of renewables and their management has proven a challenge for the prediction and control techniques developed in the last century. Since AI and ML techniques have shown promising ability to cope with complex systems in other engineering domains, researchers have been investigating their applicability to power systems. In this chapter, the application of the AI techniques discussed in Chapter 6 to energy systems is presented.

An important aspect of renewables is that, unlike thermal generators, their power output is variable depending on weather conditions. As a consequence, forecasting of power output becomes a key function both prior to construction and during operation. In Sections 7.1 and 7.2, various types of wind and solar forecasting techniques are reviewed. Algorithms based on the physical design of the systems, on statistical techniques, and on AI and ML techniques have all been used with various types of data sets. Section 7.3 describes how AI has been applied to predictive maintenance since maintenance is an important function for keeping the grid reliable. Another important function for running the grid is load prediction over both short and long time periods, and control to ensure that load and supply are balanced in real time. In Section 7.4, AI techniques applied to load forecasting are examined while Section 7.5 describes the application of AI techniques to grid control. Finally, grid security using AI is discussed in Section 7.6 extending the basic security discussion of Chapter 4, while Section 7.7 summarizes the chapter.

U. Cali et al., *Digitalization of Power Markets and Systems Using Energy Informatics*, https://doi.org/10.1007/978-3-030-83301-5_7

7.1   Wind Power Forecasting

The volatile and highly weather-dependent nature of wind power generation raises challenges in grid operation and market integration. Wind power forecasting systems provide a solution to help mitigate negative impacts of the fluctuating wind power resource. Over the past two decades, a large body of research has been conducted on wind power forecasting (WPF) [1–9]. Figure 7.1 illustrates the functional architecture of a WPF system. Two types of input data are used in WPF, static parameters and dynamic data. Parameters of the wind turbines and geographical data on the location such as power curves, the installed capacity of the wind farm, terrain information, and the coordinates of the wind turbines are the most important static input parameters. Dynamic data include weather-related data such as historical or actual numerical weather predictions (NWP), measured wind power, meteorological measurements from meteorological towers or stations, and offshore forecasting and measurement time series if the model is designed for the offshore environment.

The second layer of the system is the preprocessing layer. Depending on the WPF model, some preliminary data cleansing, plausibility for missing or faulty data, and feature engineering procedures are carried out on the input data in this stage. Feature engineering is necessary to transfer the raw or semi-raw data into a feature vector to increase the efficiency of the prediction system. This method can boost the performance of AI models. Correlation analysis or principal component analysis (PCA) are commonly used to determine best-fitting input parameters. PCA is among the most successful preprocessing techniques and uses an orthogonal transformation to convert a large set of input variables into a more meaningful

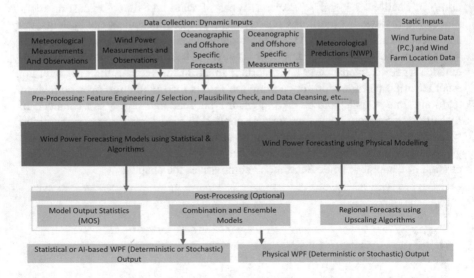

Fig. 7.1 Functional architecture of a wind power forecasting systems. (Adapted from [5])

pattern with stronger correlations. This enables a reduction in computation time necessary to reach higher accuracy.

The third layer is the wind power forecasting layer and is the core of the architecture where various types of WPF models such as physical or statistical models are deployed. Optionally, a postprocessing layer may be required depending on the WPF model. Postprocessing may include the use of ensemble models to merge various type of WPF time series or to combine forecasts, model output statistics (MOS) to reduce forecasting uncertainty, and special upscaling algorithms for regional WPF.

7.1.1 Input Data for Wind Power Forecasting Systems

Meteorological parameters such as numerical weather predictions and wind measurements, wind power output data from the SCADA system of the wind farm, and oceanographic data such as wave forecasts and measurements from offshore wind power prediction systems are the most influential dynamic factors that affect the accuracy of WPF systems. In addition, the coordinates of wind turbines, power curves, wind farm layout, and site characteristics are static factors that have an effect on the forecast accuracy.

Numerical Weather Prediction Numerical weather prediction (NWP), power curve, and measured wind power value are some of the effective prediction data for WPF models. The data sets include predicted metrological parameters such as wind speed, wind direction at various heights such as 10, 50, 100 and 130 m, temperature, momentum flux, and pressure. Depending on the meteorology service provider and their modeling structure, NWP time series can be generated once or twice a day with two to several days forecast horizons.

Wind Power Measurement Data Sets Commercial wind turbines are advanced electromechanical systems that are controlled by SCADA systems where critical operational data are also stored. SCADA systems measure the active power output of each individual wind turbine and of the entire wind farm. Measured power output is an essential input parameter especially for the statistical and AI-based wind power forecasting models since it allows the models to correlate meteorological data with the wind turbine output to achieve accurate forecasting.

Power Curve A power curve represents the power generation characteristics versus various wind speed values. Each wind turbine has its own power curve depending on the manufacturer and the technical configuration of wind turbine. The power curves are determined by using field measurements where an anemometer is mounted on a measurement mast close to the wind turbine and readings are taken for various wind speeds. In general, commercial wind turbines start generating power at 3–4 m/sec, called the cut-in wind speed. At lower speeds, there is not sufficient torque on the wind turbine blades to overcome frictional forces. The rated power indicates the installed capacity of the wind turbine when the wind turbine is operated in the rated output wind speed, typically 12–18 m/sec (see Fig. 7.2). After reaching the rated output wind speed, the wind turbine is able to generate power with the maximum capacity until the wind speed reaches 25 m/sec, the cut-out wind speed, at which point the wind turbine shuts down gradually to prevent mechanical damage.

Wind Measurements Wind information measured by anemometers can be used as an additional input parameter for statistical and AI-based wind power forecasting models. This information is generally used to forecast horizons ranging from a couple of minutes to a couple of hours. Wind speed and direction are the most important wind measurement parameters. The wind measurement equipment can be mounted on special wind measurement masts that are located in the vicinity of the wind farm. Wind turbines themselves have meteorological measurement instrumentation equipment mounted on their nacelle. However, the reliability of nacelle-mounted wind measurements is lower due to distortion of observed wind speed values. Measured meteorological parameters are useful only for short-term WPF models having maximum forecast horizon of 8 hours.

Offshore Measurements and Wave Forecasting Models Even though most wind farms are currently located onshore wind, offshore wind farm deployments are increasing. Although offshore wind farms are more difficult and expensive to build and maintain, the difficulty and expense are more than compensated by the superior

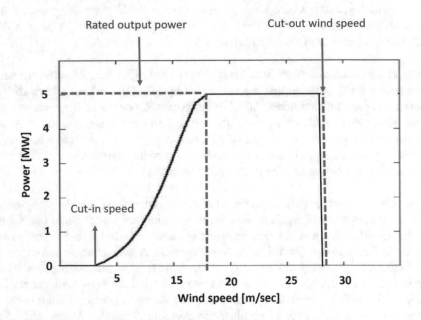

Fig. 7.2 The power curve of a sample 5 MW wind turbine

wind resource available in offshore locations. Wind resources offshore tend to be less variable and stronger than in most onshore locations.

Offshore-specific observations and measurements are the most influential input parameters for offshore wind power forecasting (OWPF) models. The Waverider buoy is a commonly used instrument to measure the ocean waves. Ocean wave measurements capture the following data:

- Spectral peak wave period
- Spectral estimate on mean wave period
- Significant wave height
- Maximum wave height
- Wavelength

Optimally radar and satellite remote sensing systems can also be utilized to observe waves and tidal currents. Use of operational wave forecasting models from various providers like the European Centre for Medium-Range Weather Forecasts (ECMWF) provide an additional input data set to dramatically increase the accuracy of predictions. The following operational wave forecasting parameters can be used in OWPF models:

- Mean wave direction
- Wave height
- Wave speed
- Wave spectral analysis
- Coefficient of drag for waves
- Wave spectral directional width
- Mean wave period of the various moments
- Mean wave period
- Directional width for swell

The observed or forecast wave parameters that correlate best with the most important WPF input parameter (i.e., wind speed) can then be chosen as an additional input to the OWPF models.

Wind–wave interaction can be understood as a means for the transfer of momentum, heat, and tidal currents between the atmosphere and the ocean. The wind at offshore and onshore locations exhibits different characteristics. The difference can be explained by investigating additional physical interactions such as air flow and thermal stratification between the sea surface and the air–sea interface. Thermal stratification is associated with water depth, incoming heat, and the temperature of the water column. Thermal stratification can be stable or unstable depending on the physical conditions. Moreover, the roughness factor of the sea surface is also dynamic and depends on the characteristics of the sea surface such as the wave amplitude and frequency. The influence of thermal stratification stability on offshore and onshore conditions has been investigated in a number of studies [3–5]. An effective air–sea interaction model for computing offshore wind profiles is presented in reference [6]. The proposed model is named an inertially coupled wind profile and is based on inertial coupling of the wave field to a wave boundary layer with constant wind shear stress.

Proper modeling of the roughness and thermal stratification variability of the sea surface are important to offshore wind power forecasting. Wave spectrum analysis is one of the methods that is used to determine the amount of energy carried by components of different frequencies in irregular sea waves. In general, wave spectral analysis is important for analyzing the effects of waves on the foundations of the wind turbine facility. This analysis can also be used to investigate the effects of the sea surface on the power output of offshore wind turbines. Statistical wind power forecasting models can tolerate a lack of detailed air–sea modeling and wave spectrum analysis if there is enough historical measured data available, although the physical parameters of the offshore environment should be considered if available. Additional offshore-specific parameters such as wave measurements and forecasts increase the accuracy of these forecasting models.

7.1.2   Wind Power Forecast Modeling Approaches

There are four main types of wind power forecasting models: physical, statistical, AI-based, and ensemble or hybrid models. The majority of forecasting systems use a hybrid approach where various forecasting models operate in parallel to decrease the uncertainty of the prediction.

Physical Wind Power Forecasting Physical WPF models are used to transform available wind speed profiles by optimizing the wind speed prediction at the hub height level of the wind turbine. Most of the classical physical models consider additional factors such as wake effects, layout of the wind farm, and power curves. It is also important to model the thermal stability of the atmosphere in addition to accounting for terrain information such as coastal orography and surface roughness. Physical models require no or less historical measured data than other approaches. Complete physical modeling where all the necessary physical parameters are taken into account is necessary to achieve accurate forecast results. In addition, the resolution of the NWP data set and terrain characteristics have significant effects on the accuracy of the physical WPF model. In general, physical WPF models require more computational power compared to statistical WPF models. In addition, poor availability of measured meteorological data to confirm NWP outputs reduces the forecast accuracy for wind farm locations [10, 11].

References [1, 2] describe early versions of physical WPF models using NWP and local meteorological measurement data sets and were developed by the Electric Power Research Institute (EPRI) in the US and Risø National Laboratory in Denmark. Many other physical models have been developed, such as the Wind Power Prediction Tool (WPPT) from Denmark [7], Previento [8] from Germany, and Zephyr [9]. Garrad Hassan (GH) Forecaster in the UK, eWind and PowerSight from the US are other commercial hybrid WPF tools commonly used by the power industry.

Existing onshore WPF models need considerable modification to adapt to offshore sites. The atmospheric and environmental conditions of onshore sites differ from their offshore counterparts. Offshore wind power forecasting models require more detailed modeling of offshore wind speed profiles considering variable sea surface roughness, thermal stratification, influence of the land–sea discontinuity, land mass, and wake effects. The impact of thermal effects and sea surface roughness on offshore wind conditions has been explored by various authors [11, 12].

Figure 7.3 shows the functional architecture diagram from Fig. 7.1 annotated to show what functions are important for onshore and offshore models. In addition to various functions such as wind–wave interaction modeling that is only needed for offshore sites, generation of local wind speed profiles normalized on the hub height levels of the wind turbines and modeling of wake effects, static and dynamic system variables are used as the input data set. MOS and upscaling functions are optional. Upscaling functions might be necessary if several wind farm clusters are being sited and regional offshore WPF time series are available. WPF time series can be deterministic or probabilistic depending on the need of the WPF end user.

Statistical Wind Power Forecasting Statistical WPF models need a sufficient amount of historically measured data to find the relationship between measured wind power and NWP to generate accurate forecasts. In general, statistical WPF models convert the NWP forecasts to a wind power prediction time series using statistical methods such as autoregression (AR), autoregression moving average (ARMA), and Kalman filters. In reference [15], the authors describe the development of the first statistical WPF based on an AR approach. Wind power output is

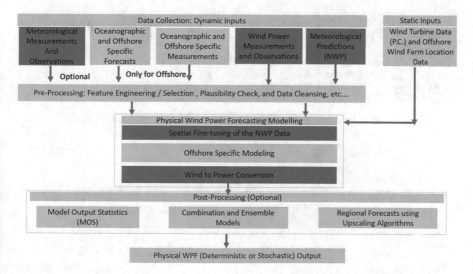

Fig. 7.3 Annotated functional architecture of a physical wind power forecasting systems. (Adapted from [5])

predicted by combining a transformation function based on the Gaussian distribution of wind speeds, the measured power curve, and a wind speed as input to a transform function based on a power law. References [16–18] describe the use of ARMA and autoregressive integrated moving average (ARIMA) models to predict and simulate wind speed time series. Kalman filters are an alternative approach to predict wind speeds. In [19, 20], the authors demonstrated that use of Kalman filtering not only improves the forecast accuracy but also the WPF results. Usually, at least one year of historical data is sufficient to catch the seasonality effect, but, in some cases, 3 months of data may deliver acceptable results. Statistical WPF models are cheaper and easier to develop in comparison to other WPF models. Figure 7.4 demonstrates an overview of the statistical wind power forecasting functional architecture, annotated with optional and offshore steps.

AI-Based Wind Power Forecasting   Similar to the statistical WPF models, AI-based models also derive a relation between the dynamic and static input parameters such as NWP and measured wind power time series from the past to forecast wind power output for the future. However, AI-based WPF models view the problem as being non-linear and utilize AI and machine learning (ML) algorithms instead of clearly defined, basically linear statistical equations. The main advantage of AI-based WPF models is their capability to predict the wind power output without knowing the physical conditions of the wind farm. Thus, detailed physical modeling is not required, although if physical information is available it can increase the accuracy of the AI model. A variety of AI and ML algorithms have been used to predict wind power output:

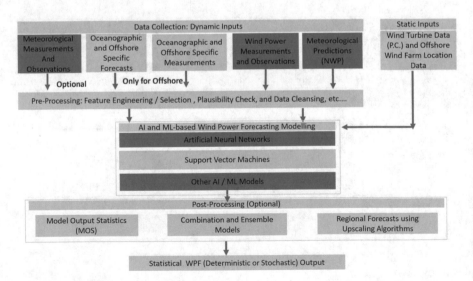

Fig. 7.4 Annotated functional architecture of statistical wind power forecasting system. (Adapted from [5])

- Artificial neural networks (ANNs) [3, 19, 20]
- Support vector machines (SVMs) [21]
- Neuro-fuzzy networks (NFN) [22]

Figure 7.5 shows a functional architecture of an AI-based WPF system again annotated with optional and offshore steps.

Most of the existing work is for onshore WPF and few WPF models have been derived to date for offshore environments. FNN-based approaches have been demonstrated for an offshore wind farm with an installed capacity of 5 MW in Tunø Knob, Denmark. The FNN-based model was developed by ARMINES and delivered similar or even better results than the onshore WPF models in terms of performance and accuracy [23]. An adaptive Markov-switching autoregressive model for offshore wind power forecasting with a new parameterization of the model coefficients has yielded better results in comparison to autoregressive model-based forecasts [13].

Ensemble and Hybrid Wind Power Forecasting   Hybrid models usually combine multiple physical and statistical models and/or NWP data sets of various origins. This approach aims to exploit the best available predictions from multiple models to reach the highest available accuracy. Hybrid models increase redundancy but may require more computation time to generate the combined prediction output. The ANEMOS project is one of the most successful implementations of a hybrid approach for both onshore and offshore sites [10, 24]. An ensemble-based probabilistic offshore wind forecasting approach has been introduced for another Danish offshore wind farm Horns Rev with an installed capacity of 160 MW [25]. The proposed model employed local polynomial regression and orthogonal fitting

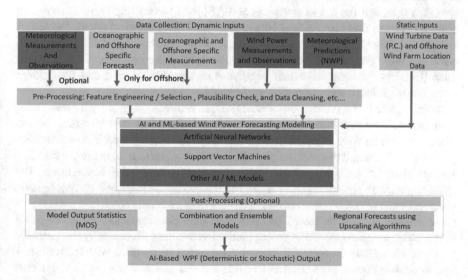

Fig. 7.5 Annotated functional architecture of AI-based wind power forecasting systems. (Adapted from [5])

methods to increase forecast accuracy. Another novel offshore WPF approach has been developed using additional oceanographic parameters such as wave measurements and forecasts [3]. A hybrid model was proposed using various statistical forecasting methods and an ensemble NWP data set that increased the forecast accuracy. A wake-adjusted physical power model and offshore-specific WPF models were demonstrated in [26].

7.2   Solar Power Forecasting

Like wind energy, solar energy has a noncontrollable and fluctuating power generation pattern. Power transmission and distribution companies need to manage the intermittent power generation from solar power-generating facilities. The use of accurate solar power prediction systems to manage an increasing amount of solar power on existing power grids is a critical management task for utilities. Utilities use solar power predictions to manage their operations and keep their power system stable just as they use wind power predictions. Determining the ramping events, optimization of unit commitment, and power transmission scheduling are mission critical daily tasks for power system operators where renewable energy forecasting models are integrated into the system. The performance of solar power is measured using statistical metrics such as normalized root mean square error (nRMSE), mean absolute percentage error (MAPE), or mean absolute error (MEA).

Solar power and irradiance forecasting models basically use the same underlying approaches as wind power and speed forecasting models. Forecast horizons, type of input parameters such as solar power measurement time series, meteorological forecast data sets, and the use of statistical and AI-based forecasting algorithms are all similar between these two energy forecasting domains. However, due to the different nature of the physical principles, there are some significant differences between solar and wind power forecasting models. Forecasting the global horizontal irradiance (GHI) is the most important parameter in solar forecasting and has a large influence on solar power generation predictions.

Solar power is the conversion of energy from sunlight using solar thermal, concentrated solar power (CSP), or photovoltaic (PV) systems into electrical power. The energy conversion principle of CSP and solar thermal systems is based on thermodynamic rules and thermal energy is harvested from the sun in the first step. CSP systems then convert the generated heat to electrical power using heat engines. PV systems generate electrical power using the photovoltaic effect. PV systems generate direct current (DC) power, which varies depending on the sunlight's intensity or solar irradiance. The harvested DC power is then converted to AC power using power electronic devices. Solar power prediction systems are basically the same for CSP and PV systems, except CSP systems can often be coupled with thermal storage to enable power generation even after the sun has set, whereas PV systems require batteries for storage.

7.2.1   Input Data for Solar Forecasting Systems

Solar forecasting uses a variety of input data types. The following are the most important input data types for solar forecasting.

NWP NWP is the most common input data set for solar power forecasting (SPF) systems. NWP models are operated by public or private weather services to predict the weather conditions in their coverage area or globally. They predict upcoming weather by using physical laws and prognostic equations such as Navier–Stokes equations [27]. Global models predict the atmospheric status of the entire Earth. Global NWP models are generated by a limited number of providers such as the Global Forecast System (GFS) from the US National Oceanic and Atmospheric Administration (NOAA) and the Integrated Forecast System (IFS) from ECMWF. Data assimilation models are used to integrate the observed meteorological information from the meteorological stations as well as ad hoc measurement systems to increase the accuracy of the NWP. Mesoscale models are then calculated based on the global models to predict the atmospheric status of specific locations on the Earth. The most important NWP data for SPF models are solar radiation density, solar irradiation, cloud coverage, temperature, humidity, pressure, and wind speed. Irradiance prediction time series can be generated based on NWP. Many NWP models can predict surface solar irradiance, which can then be directly used as an input parameter to an SPF system. The spatial resolution of the NWP can vary depending on the local model and usually ranges between 27 km × 27 km down to 1 km × 1 km [28].

Solar Irradiance Solar irradiance is a measure of the visible and infrared light power per unit area received from the sun. Productivity of solar power systems is mainly dependent on this parameter. Solar irradiation can be categorized into four groups: direct, diffuse, total, and global irradiance. Direct solar irradiance or direct normal irradiance (DNI) is a measure that represents the amount of solar beam radiation received perpendicular to a surface (such as a PV panel). Diffuse horizontal irradiance (DHI) is a measure of solar radiation received following an indirect path from the sun and passing through some atmospheric obstacles such as air, cloud, and aerosol particles. Total solar irradiance (TSI) is a measure of the solar power that indicates the totality of solar energy as a function of wavelength per unit area incident arriving to the Earth's upper atmosphere. Global horizontal irradiance (GHI) is the total solar radiation received from above by a surface horizontal to the ground. In other words, it is the sum of DHI and DNI by considering the cosine of the zenith angle.

Power Output Data Solar power measurements can be provided by data acquisition and metering devices in small-scale solar PV installations or from SCADA systems in larger solar PV farms.

Sky Imager Data Sky imagers are used to detect cloud movement. Use of sky imagers for short-term solar power forecasting can significantly increase the accuracy of the predictions. Sky imagers are very efficient at capturing sudden changes in solar irradiance and ramping effects in solar power output. According to [29], the most effective time horizons for sky imagers are between 5 and 30 min.

Satellite Data Satellites can provide information from geostationary orbit. In particular, dynamic cloud movement information and surface irradiance can be delivered from satellite data [30]. In reference [31], the authors successfully applied an operational satellite-based irradiance forecasting system to SPF models.

7.2.2   Solar Power Forecasting Model Approaches

Just as with wind power forecasting, there are four main approaches to solar forecasting: physical methods, statistical methods, artificial intelligence methods that use irradiance data, and ensemble and hybrid methods that combine two or more of the other approaches.

Physical Methods The physical SPF models are designed to convert GHI or solar irradiance and other meteorological variables into predicted power output. Physical forecasting models are based on NWP, total sky imager measurements, cloud observations delivered by satellites, and other relevant measurements. Predicted solar irradiation is the most important variable for physical SPF models. Static parameters, such as the location of the PV farm, PV configuration including tilt angles and installed capacities, are combined with dynamic data, such as NWP and meteorological observations, from sources like satellite images and global sky imagers, and are input into the physical SPF modeling algorithms. The spatial refinement and solar energy conversion-related equations are deployed to generate the predicted solar power.

The output predictions depend on the forecast horizon and spatiotemporal resolution. Depending on the methodology, deterministic or probabilistic SPF time series can also be generated. Multiple researchers have developed physical SPF using different methodologies [32–37]. The main advantage of physical SPF models is that no historical input data is required. On the other hand, developing physical models is a time-consuming task and lack of spatiotemporal information may lead to very poor PV power prediction time series [38]. Depending on the use case, special upscaling functions can be deployed to generate the regional SPF time series. For example, in reference [35], the authors developed ensemble SPF models for spatially dispersed grid-connected PV systems. Figure 7.6 illustrates the functional architecture of a physical SPF system.

Statistical Methods Statistically generated solar models are based on historical data such as solar irradiance and power measurements to predict future PV power output. Statistical models have the ability to correct systematic errors. AR [39],

ARMA [40], ARIMA [41], and coupled autoregression and dynamical systems (CARDS) [42] are commonly used statistical methods in solar forecasting. Seasonal variability has been integrated into the statistical models using seasonal ARIMA (SARIMA) [43]. In reference [44], an SPF model is demonstrated based on a probabilistic vector AR (VAR) approach.

Artificial Intelligence Solar Power and Irradiance Foresting Models   Like statistical SPF models, AI-based models also strongly rely on historical dynamic data. ANNs are the most common AI-based SPF models and use various machine learning configurations such as feedforward, recurrent wavelet, and radial basis function neural networks [45–47]. The ANN consists of input, hidden, and output layers with links (called neurons) between them. The input data is inserted into the input layer. The hidden layers process the given input data and apply various functions, then forward the processed data to the output layer where the solar power predictions are generated. The neurons have two main functions: combination and activation. The neurons are responsible for linking the various layers and for adjusting the dynamically updated weights helping the entire system learn from the historical data. The performance of an SPF model is strongly correlated with the design of the ANN architecture.

When building an ANN model, the available data set is split into a training and a testing data set. In some cases, a validation data set is also incorporated to the model development procedure. The training data set allows the learning algorithm to learn the relationship between the input parameters (mainly NWP) and the solar power measurements are from the historical data, mainly by adjusting the synaptic weights on the neurons. The deviation between target input and output is calculated iteratively. This process minimizes the error between the input and output, but care must be taken to not overfit the model to the input data set; otherwise, it has little

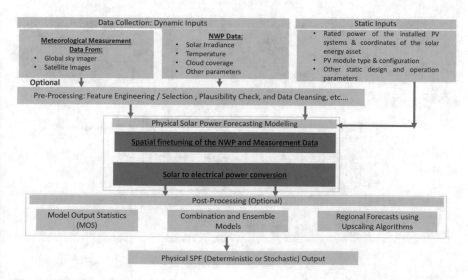

Fig. 7.6 Functional architecture of physical solar power forecasting systems. (Adapted from [5])

predictive value beyond the data set used to generate it. It is also possible to deploy more advanced NN models such as the adaptive neuro fuzzy interference system (ANFIS), which is based on the Takagi–Sugeno fuzzy interference system principle. ANFIS SPF models combine ANN and fuzzy logic approaches where a set of IF-THEN procedures are integrated to the model to approximate the nonlinear functions [48] underlying the solar power generation system. Other AI approaches including support vector machines (SVM) [49], random forest, and decision trees [50] have been used to develop AI-based SPF models. Figure 7.7 contains the functional architecture of statistical and AI-based SPF systems.

Ensemble and Hybrid Solar Power Forecasting Hybrid models reduce prediction errors and increase redundancy by incorporating multiple approaches into one prediction model. Hybrid models are applied in either the pre- or postprocessing stage. Applying feature engineering and optimization techniques such as principal component analysis (PCA), genetic algorithms (GAs), wavelet transform (WT), SVMs, and particle swarm optimization (PSO) has the potential to increase forecast accuracy and quality through preprocessing. Hybrid approaches have also been developed using GA + WT + PSO [51] and GA + SVM [74], SARMA and SVM (2013) [43], and SVM and least square (LS) [52].

7.3 Predictive Maintenance

Predictive maintenance (PDM) is a technique using data analysis tools to detect anomalies in a system that predict that a component may soon require maintenance before the component fails so that maintenance can be conducted [53]. PDM is

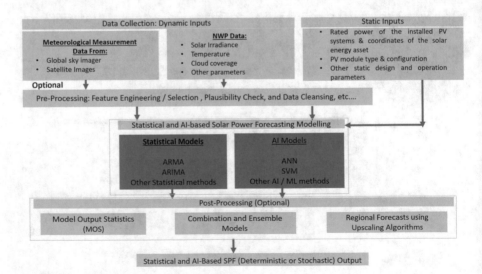

Fig. 7.7 Functional architecture of statistical and AI-based solar power forecasting systems. (Adapted from [5])

sometimes called condition-based maintenance (CBM), but PDM typically includes data measured by IoT sensors. The traditional approach to maintenance is to conduct maintenance based either on a fixed schedule or waiting for the machine or system to fail. These approaches are sometimes called time-based maintenance (TBM), failure finding maintenance (FFM), and corrective maintenance. Predictive maintenance, on the other hand, is based on an asset's actual state measured by sensor devices.

Maintenance is an essential requirement to support the reliability of the electric grid. Electric power generation systems include many types of generators, diesel engines, gas turbines, step-up and step-down power transformers, motors, compressors, pumps, switchgear equipment, etc. The transmission system includes overhead lines and underground cables and equipment to ensure self-monitoring and self-healing. The distribution system includes smart meters, switching gear, cables, fuses, isolators, circuit breakers, relays, control panels in addition to energy storage devices, such as lithium batteries, flow batteries, compressed air, super-conducting magnetic energy storage, super-capacitors, and flywheels. These assets need to be managed for reliability and optimum economic performance over their lifetimes.

IoT measurements supply data real-time data that is then combined with data analytics techniques to allow the identification of when an asset will require maintenance and thereby help prevent equipment failure. Measurements can be stored as reference information for later comparison. Predictive maintenance offers utilities the least expensive, most efficient method of reducing equipment-related down time, and is widely used to improve the reliability, efficiency, and maintenance cost, as well to help ensure the safety of the electric power grid. Although there are many PDM solutions available, most industrial companies still use regularly scheduled maintenance rather than analytics tools. Using nonautomated data analytics requires people to analyze collected data, evaluate the results and determine when maintenance is required. This approach is labor intensive, not scalable and depends heavily on the team involved [54].

Research in automated predictive maintenance mainly focuses on the insulation materials of different electric grid components, generators, power transformers, underground power cables, circuit breaker, switching gear, relays, and motors [55]. When a large quantity of data is collected from monitoring, the useful information about defects and their severity may be hard to interpret. Different types of diagnostic techniques may be employed. Online diagnosis is used to provide a quick overview in case of material degradation. Offline diagnosis is used periodically to determine the exact cause and location of any pending defects using trend analysis and AI. The analysis and diagnostic stage have several substages including early detection and diagnosis, isolation, fault identifiability, robustness, novelty identifiability, explanation facility, and adaptability. A variety of conditions can be monitored based on the electrical equipment. For a transformer, these include the temperature, SF6 concentration, insulation breakdown; for the condition of a tap changer, the operating time, circuit breaker opening and closing time, and heat; for switch gear, the temperature of connections, and the insulation condition; for a motor, the starting time, number of operations, and power, insulation breakdown; and forcible connections, insulation breakdown [56].

Predictive analytics involves building a big data framework to monitor asset performance through continuous collection and analysis of sensor data. The analysis provides in-advance warnings of component failures. The amount of data from different sensors and other parts of subsystems is huge; hence, a big data framework coupled with machine learning can be implemented to solve this problem. Developing an effective predictive maintenance solution requires machine learning models built on clean data, information collected from a multitude of sensors, and robust historical records that encompass a thorough characterization of the equipment. Deployment success is based upon algorithm performance that can manage real-time data flow and ingestion. A variety of machine learning methods have been applied to the predictive maintenance, including logistic regression, decision trees, K-nearest neighbors (KNN), support vector machines (SVM), artificial neural networks (ANN), deep learning, K-means, and naive Bayes, and in many cases combined into hybrid techniques. Table 7.1 summarizes various studies using machine learning for predictive maintenance.

7.4   Load and Net Load Forecasting

Electricity load forecasting is important for the economic operation of power systems. Utilities use load forecasting to make both short-term and long-term decisions about how to meet future load demands [65]. Similarly, consumers can use load

Table 7.1 The summary of machine learning methods for predictive maintenance

System/device	Algorithm	Sensor/data
Nuclear power plant, Turbofan Engine [57]	LR[a], SVM	Lidar sensors
Medium voltage switchgear [58]	SVM	Temperature, breaker drive monitoring, partial discharge
Photovoltaic (PV) system [59]	ANN	Solar irradiance, PV panel temperature
Wind turbine Gearbox [60]	ANN	SCADA data regarding wind turbine subcomponents/sensors
SF6 circuit breaker [61]	Fuzzy logic, K-means, and ANN	State of contacts, the chamber, the SF6 medium and the motor-driven mechanism
High-voltage equipment [62]	Fuzzy logic	Number of disturbances, failures of pressure and density sensors as well as alarm switches)
Power transformer [63]	ANN, fuzzy logic and expert systems	The oil chromatographic data
Photovoltaic (PV) system [64]	CNN	Daily power data

[a]Logistic regression algorithm (LR)

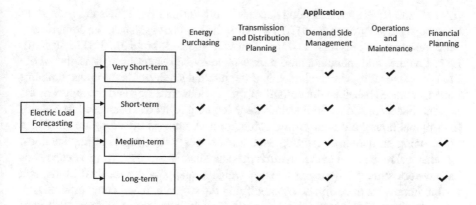

Fig. 7.8 Classification of load forecasting techniques based on forecasting interval

forecasts to reduce energy consumption [66]. Load forecasts must be accurate both in terms of magnitude and geographical location over different time horizons.

Load forecasting techniques can be classified based on the time horizon they model as shown in Fig. 7.8 [67]. Very short-term electric load forecasting techniques range from a few minutes to an hour, while short-term forecasting ranges from hours to weeks. Very short-term and short-term forecasting are used for energy purchasing, transmission and distribution planning, demand side management, and scheduling of maintenance. Medium-term forecasting ranges from 2 weeks to 3 years, while long-term forecasting ranges from 3 years to 50 years. These techniques are used by utilities for energy purchasing, demand side management, infrastructure upgrade planning, and financial planning.

Traditional load forecasting methods include statistical models such as time series and regression. Time-series models decompose the historical data over equal intervals of time into three components: trend, seasonality, and residual. These methods use the decomposed data to identify underlying patterns and form a model to forecast into the future [68]. Some common time-series methods include AR, ARMA, and ARIMA. Regression methods are the most commonly used technique for load forecasting since they are relatively simple to implement. For load forecasting, factors such as metrological effects and day types are used as the independent variable [69]. Regression techniques can be extended to use multiple independent variables for forecasting [65]. For example, the effect of temperature on load has been demonstrated to depend on geographical location using multiple regression [70]. Exponential smoothing methods use weighted averages of past observations with weights decaying exponentially [69].

ML and AI models have been used when the relation between load and available data is nonlinear. Though some traditional methods are good at modeling nonlinearity, AI models have been shown to be better for load forecasting in some cases [71–73]. However, AI models are complicated and hard to interpret. In the recent years, some work has shown that a combination of different models provides better results. For short-term load forecasts, a combination of long short-term memory

(LSTM) and RNNs has had good success in prediction [74]. RNNs are trained by backpropagation through time (BPTT) but if the gradient vanishes, the components decrease exponentially fast to zero. To overcome this problem, LSTM is used. LSTM defines and maintains an internal memory cell state throughout the whole life cycle, indicating which elements of the internal state vector are updated, maintained, or erased based on the outputs of the previous time step and the inputs of the present time step. Along similar lines deep learning RNN has been used to classify the original household load profile using spectral analysis offsetting uncertainties [75]. Doing so avoids overfitting that often arises with deep learning methods. Another technique used in the research literature involves constructing a deep residual network structure by using a set of weights, mapping the residual block, and further improving the estimate by modifying the structure using convolutional network structures [76]. Monte Carlo dropout probabilistic forecasting was also used to capture the uncertainty within this model.

Some researchers have considered combining renewable energy with energy storage systems such as batteries [77]. The societal goal of reducing greenhouse gas emissions has resulted in incentives for consumers to purchase distributed photovoltaic (PV) (rooftop PV). Because these systems are typically built behind the meter, they are not visible to utility forecasting systems and they introduce challenges in obtaining reliable load forecasts [78–80]. With development of technologies to determine building load and rooftop PV output, net load forecasting performance can be improved at the aggregate level. Other issues arise in load forecasting due to uncertainties in the data, models, and land use decision. Human behavior, climate variability, and intermittent power generation from renewables can cause uncertainty in the collected data. They add stochastic uncertainty to net load forecasts at aggregated levels. Some research studies show that these uncertainties can be minimized by using probabilistic forecasting techniques. Probabilistic forecasting techniques provide interval bounds but may have residual error requiring post processing. Land use decisions due to new urban development can cause additional uncertainty. Planners using historical load data can miss loads from such new development. Finally, uncertainty in models arises because even though a large number of variables have been explored to minimize the model uncertainty, there still exists uncertainty in the model parameters and the model structure themselves.

Load forecast accuracy is further impacted by unanticipated beneficial electrification infrastructure developments where electric infrastructure is deployed within the distribution grid to get rid of fossil carbon generating energy infrastructure. GHG reduction policies have promoted the addition of electric vehicles (EV) to both the personal transport, commercial transport, and public transport sectors. EV charging behaviors are stochastic in nature and hard to predict. With increase in renewable energy sources, local energy storage facilities such as batteries, hydrogen storage and flywheels are being deployed both in front of the meter as well as behind the meter. These devices can have a major impact on load variability.

Bayesian deep learning techniques (Bayesian deep LSTM neural networks) have been used to minimize uncertainty from behind the meter PV deployments, human behavior, and uncertainty in models by deploying clustering technique to enhance the aggregate forecasts in probabilistic forecasts [77] and hyperparameter tuning [78]. A data-driven approach has also been used to address these uncertainties in short-term forecasts [79].

7.5   Power System Control

With the increasing penetration of renewable DERs, research into decentralized and distributed computational algorithms for power system control has become necessary. Rooftop PV systems, battery energy storage systems (BESS), demand response (DR) programs, and EVs have transformed residential, commercial, and industrial consumers to prosumers. In such a scenario, the behavior of prosumers and uncertainty in renewable power generation can increase the variability of the load versus supply balance. This transformation has led the research community to consider power control as stochastic. The dynamic nature of power systems with a large numbers of DERs raises the complexity of load–frequency control, voltage control, and power system protection and security. Moreover, the market trend in power systems has created smart and organized market players with organized strategies other than the traditional market players (generators, TSO, and DSO). These trends have led researchers to consider how AI and ML can be used for power systems control.

7.5.1   Artificial Intelligence Algorithms Investigated in Power Control Research

CNNs and RNNs are the most frequently used algorithms in power control research, where CNN has been used for handling spatial distribution data and RNN for temporal data [81]. ML techniques include supervised, unsupervised, semisupervised, and reinforcement learning (RL)-based methods. Supervised learning needs to be trained with a labeled data set for building an agent while unsupervised learning provides only unlabeled data to the algorithm to identify clusters or groups in the data. Semisupervised learning uses labeled and unlabeled data sets for model training. RL includes an algorithm for feedback to learn and predict. The components in RL are agent, environment, reward, and action. An agent in RL maximizes a reward by following a series of actions responding to a dynamic environment. Q-learning, state action reward state action (SARSA), deep Q-networks (DQN), and deep deterministic policy gradients (DDPG) are extensively used algorithms in RL. Deep research learning (DRL) is the fusion of DL's perception and RL's decision-making [82]. Deep learning (DL), reinforcement learning (RL), and deep reinforcement learning (DRL) are the most frequent ML algorithms used in smart grid control research.

7.5.2   Traditional Power System Control

Power system control has traditionally been accomplished by using automated generator control (AGC) systems to balance the power output of multiple generators at different power plants in response to changes in load in order to balance the supply and demand of power. AGC systems maintain the system frequency by adjusting the active power. The performance of AGC is measured by frequency deviation, area control error (ACE), and control performance standard (CPS) from the control parameters. A flywheel governor on a synchronous machine was the first attempt at AGC. Ancillary control with a feedback control loop dependent on frequency deviation was provided to the governor since the technique was found to be inefficient. The initial literature published on AGC was based on tie-line bias control. Follow-up work explored noninteractive control of frequency and tie-line power controls. The performance standards related to AGC have been continuously updated as the complexity of the power grid has grown. Also, most of the initial work modeled AGC using a linearized model of multiarea power systems. Nonlinearity was included in the models that followed. Since smart grid-based control and security involve handling large amounts of data, uncertainty, and state estimation, the computational efficiency of conventional approaches may not be sufficient for optimal control [83].

7.5.3   AI/ML for Power Systems Control

One major problem posed by the increase in renewable energy sources is the reduction in system inertia. This has caused increased challenges in maintaining balance since unlike spinning thermal generators renewables inherently don't have any inertia. On the other hand, the variations of renewable power generation create new challenges to voltage control. Research has been conducted to utilize RL and DRL for AGC and voltage control. Table 7.2 summarizes machine learning applications in power system control in the literature.

7.5.4   Demand Response

Demand response (DR) programs are an especially promising feature that the modern distribution network can use for distribution systems control. DR is primarily classified into price-based and incentive-based DR schemes. Price-based DR schemes control the load by changing the price of electricity over different time periods. Schemes like time of use (ToU), critical peak price, and real-time price fall into this category. Incentive-based schemes provide incentives to the end-use customers for reduction in consumption. Direct-load controls, demand-bidding programs, and interruptible tariffs fall under incentive-based DR programs. Identifying

Table 7.2 Summary of machine learning models for power system control

Ref.	Application	Data set	Model	Performance metric
[84]	Stochastic optimal AGC	China Southern Power Grid (CSG) case	Multi-step Q learning (RL)	Frequency deviation $|\Delta F|$, area control error $|ACE|$, control performance standard (CPS) (%)
[85]	Multiagent smart generation control	Two-area IEEE power system, CSG	Decentralized correlated equilibrium Q(λ)-learning algorithm	$|\Delta F|$, $|ACE|$, CPS(%)
[86]	Multiagent smart generation control	Two-area IEEE power system, CSG power system model	Decentralized win or learn fast policy hill-climbing (DWoLF–PHC (λ))	$|\Delta F|$, $|ACE|$, CPS(%)
[87]	Reactive power control	IEEE 14 busbar, IEEE 136 busbar	Q-learning (RL)	Voltage profile
[88]	Two-timescale voltage control in distribution	IEEE 123-bus test feeder, real-world 47-bus distribution system	DRL	Voltage profile
[89]	Voltage control	IEEE 13-bus and 123-bus	Markov decision process with batch RL	Voltage profile
[90]	Autonomous voltage control	Realistic 200-bus test feeder	DQN and DDPG (DRL)	Voltage control action
[91]	Voltage control	IEEE 10 machine 39 node system	Q-table (RL)	Voltage profile

and predicting the trend of the supply and demand is pivotal for implementing DR. Nonintrusive load monitoring through smart meters allows the utility or load aggregator to analyze energy consumption behaviors. AI and ML methods used in DR are reviewed in reference [92].

Forecasting is a key component for effective DR. As discussed in Section 7.4, AI has been used for load forecasting for planning purposes and can also be applied to load and price forecasting for DR. Load forecasting has been done for different aggregation levels such as residential, large building, and equipment level. Reference [93] compares ordinary least-square regression (OLS), L1 and L2 regression, k-nearest neighbors regression (KNN), support vector regression (SVR), and decision tree regression (DT) for predicting residential energy consumption. Results are heavily dependent on the data set utilized. Some research has also used ANN-based algorithms for domestic load forecasting. Baseline load estimation predicts the load without considering DR. Baseline load estimation plays a key role in identifying the customer's power consumption behavior, thus enabling the decision maker to identify what kind of DR program suits a cluster of customers. A feeder with lower PV penetration may have a load profile with peak consumption at noon while a feeder with a high PV penetration may have an evening peak load profile. Flexibility forecasting is essential in understanding and optimizing the potential of DR. A DL model factoring four-way conditional restricted Boltzmann machines (FFW-CRBMs) is utilized in reference [94] to identify and predict building energy flexibility. The paper uses a hybrid approach combining sparse smart meters and ML models to compensate for the unavailability of consumer smart meter data. Demand-side energy management is also an essential part of DR. The decision of a consumer to buy or sell energy for each time period by identifying the requirement for energy and the historical energy price is explored employing batch RL in reference [95]. Considering the impact of future energy prices and future consumer device selection, Q-learning was used for residential demand–response for energy management systems in reference [96].

7.6 Grid Security

As discussed in Chapter 4, the introduction of Internet connections to grid networks for control purposes opens the potential for attacks from outside. If grid systems are left unprotected, attackers can break into the control systems and modify control readings or cause complex equipment to malfunction. AI techniques have been applied to grid security in two areas: differentiating anomalies from attacks and intrusion detection. These topics are discussed in Sections 7.6.1 and 7.6.2, respectively. Table 7.3 summarizes AI methods reported on in the research literature with the attack types and data sets.

7.6.1 Differentiating Anomalies from Attacks Using ML

Smart grid systems are subject to various anomalies from equipment malfunctions, weather-related problems, and the like that have nothing to do with attacks. Detection methods have been developed to detect such anomalies, allowing the smart grid management systems to automatically correct the parameter settings of the malfunctioning systems for certain smart grid services. Existing anomaly detection methods include protection-based approaches and detection-based approaches. Distribution networks are especially vulnerable because they were designed to be passively managed, but due to a shift toward increasing deployment of DERs and electric vehicles (EVs), the distribution grid is undergoing structural changes requiring more reliable and real-time monitoring.

However, these detection systems, if left unprotected, can be a target for attackers. An attacker can compromise the smart grid management systems by injecting false data into the sensors, allowing compromised readings to pass through the grid management system as though they were valid. The data then passes to the management algorithms, corrupting the management system and resulting in malfunctions and equipment failures. The existing anomaly detection methods are by and large

Table 7.3 Summary of machine learning algorithms security

Attack type	Algorithm	Data set
False data injection to PMU data [97]	SVE, perceptron, SVM, k-NN, SLR[a]	Synthetic data set
Denial of service (DoS), R2L, U2R, probe, and Normal class [98]	k-NN, NN, DT, random forest	KDD99 data set, NSLKDD
Energy fraud [99]	ANN	Synthetic data set
Stealthy false data injection [100]	SVM	Synthetic data set
False data injection to smart meter data [101]	DT and SVM	Private data set
False data injection to smart meter data [101, 102]	SVM	Private data set
False data injection to smart meter and sensor data [102]	SVM and KNN	Synthetic data set
False data injection to PMU data [102]	RBM[a]	Synthetic data set
Cyberattack to grid field devices such as RTUs[a], PLCs[a], PMUs[a], and IEDs[a] [103]	SVM	Synthetic data set
False data injection to the power flow data [104]	SVM, KNN, and ANN	Synthetic data set
False data injection to SCADA system [105]	CDBN[a]	Synthetic data set
False data injection to PMU data [106]	CDBN[a]	Synthetic data set

[a]Sparse logistic regression (SLR), restricted Boltzmann machine (RBM), remote terminal unit (RTU), programmable logic circuit (PLC), phasor measurement unit (PMU), intelligent electronic devices (IED), and conditional deep belief network (CBDN)

unable to detect this kind of stealth attack called a false detection injection attack (FDIA) [97].

Data analytical methods coupled with machine learning (ML) have been employed to mitigate FDIAs, especially when large-scale smart grids generate huge amounts of data. ML utilizes training data to construct a model of the system when it is functioning properly. This model can then be used to detect anomalies and initiate control actions. With ML, the grid management software can be trained to detect FDIA. Unlike other FDIA detection algorithms, ML algorithms are based directly on data collected from the system. Many research studies have been conducted on ML cybersecurity approaches for the transmission grid, but research work on ML cybersecurity for the distribution grid is still in an early stage.

7.6.2 Intrusion Detection Systems (IDS)

Machine learning methods have also been widely used for intrusion detection systems (IDS) to detect and identify any anomaly that could stem from an unauthorized party gaining access to the network. The ML method selected is crucial for achieving acceptable performance. Logistic regression, decision trees, KNN, SVM, ANN, DL, K-means clustering, naive Bayes, and combined and hybrid methods have all been tested for IDS. ML algorithms depend on the quality of the data, and since collecting data on intrusion exploits is difficult in operating networks, most studies depend on benchmark data sets. Benchmark data sets additionally make experimental results more convincing and simplify comparing the results with those of previous studies. A common source of IDS data sets are packets, flows, sessions, logs, and data patterns obtained from public benchmark data sets. Some example public data sets are DARPA1998 [107], KDD99 [108], NSL-KDD [109], UNSW-NB15 [110], and ADFA-LD [111]. Among them, KDD99 is the most commonly used IDS benchmark in academic research. It was derived from DARPA1998, and the labels in KDD99 are the same as those in DARPA1998. Machine learning methods for IDS are trained to perform binary classification, that is, anomaly or normal, and multiclass classification to predict attack categories, such as normal, DoS (denial of service), U2R (user to root), R2L (remote to local), and Probe (probing attack) [112]. A data set can also be constructed through simulation experiments if the benchmark data sets do not provide sufficient detail.

7.7   Summary

Research into AI and ML technologies for power systems has provided some important solutions to the problems of the grid's ever-growing complexity. Wind and solar power forecasting research has been especially useful in predicting the power output from utility scale and DER-renewable energy facilities. This work will be important to extend, broaden, and bring into active service as the penetration of renewables into the grid continues in coming years in response for the societal goal of a carbon-free grid. AI and ML research has also shown promise for predictive maintenance and load prediction for both infrastructure upgrade planning and DL program planning. Similarly, research into the application of AI and ML to power system control is promising, especially given the variable nature of renewables and their impact on frequency control, Volt/Var control, and capacity. Finally, AI and ML techniques have been used for security challenges unique to the grid, such as false data injection, as well as those common to all networks, such as intrusion detection. Application of AI and ML techniques to the energy domain is still in its infancy, and over the coming years, the grid is likely to see more deployment of these powerful computational techniques.

References

1. McCarthy E. (1998). Wind Speed Forecasting in the Central California Wind Resource Area. Paper presented in the EPRI-DOE-NREL Wind Energy Forecasting Meeting, Burlingame, CA, USA.
2. Troen, I., & Landberg, L. (1990). Short-Term Prediction of Local Wind Conditions. Madrid / Spain: Proceedings of the European Community Wind Energy Conference, pp. 76–78.

3. Cali U. (2010). Grid and Market Integration of Large-Scale Wind Farms Using Advanced Wind Power Forecasting. Technical and Energy Economic Aspects. Kassel, Germany: Kassel University Press GmbH.
4. Lange B. (2002). Modelling the Marine Boundary Layer for Offshore Wind Power Utilization, PhD Thesis, 2002.
5. Cali, Umit, and Claudio Lima. "Energy informatics using the distributed ledger technology and advanced data analytics." Cases on Green Energy and Sustainable Development. IGI Global, 2020. 438–481. (2)
6. Kariniotakis, G., Halliday J., R. Brownsword, I. Marti, & A.M. Palomares (2006). Next Generation Short-Term Forecasting of Wind Power – Overview of the ANEMOS Project. Athens, Greece: European Wind Energy Conference, EWEC 2006,10 p.
7. Nielsen, T.S., H. Madsen, & J. Tøfting (1999). Experiences with Statistical Methods for Wind Power Prediction. Nice, France: Proceedings of the European Wind Energy Conference, pp. 1066–1069.
8. Focken, U., Lange M., Waldl H.-P. (2001). Previento – A Wind Power Prediction System With an Innovative Upscaling Algorithm. Proceedings of the European Wind Energy Conference: , Copenhagen, Denmark, pp. 826–829.
9. Giebel, G., Landberg L., Nielsen T.S., Madsen H. (2002). The Zephyr Project – The Next Generation Prediction System, , Paris, France: Global Wind Power Conference and Exhibition.
10. Tambke, J. (2006). Short-term Forecasting of Offshore Wind Farms Production – Developments of the Anemos Project. Athens, Greece: Proc. of the European Wind Energy Conference 2006, 27/2–2/3.
11. Tambke, J., Lange, M., Focken, U. & Heinemann, D. (2003). Previento meets Horns Rev - Short- term wind power prediction - Adaptation to offshore sites in CD. Madrid, Spain: Proceedings of the 2003 European Wind Energy Association Conference, EWEC'03.
12. Watson, S.J., & Montavon C. (2003). CFD Modelling of the Wind Climatology at a Potential Offshore Wind Farm Site, Madrid, Spain: European Wind Energy Conference EWEC
13. Pinson, P. & Madsen H. (2012). Adaptive Modelling and Forecasting of Offshore Wind Power Fluctuations with Markov-switching Autoregressive Models. Journal of Forecasting, 31(4), pp. 281–313.
14. Brown, B.G., Katz, R.W. & Murphy, A.H. (1984). Time Series Models to Simulate and Forecast Wind Speed and Wind Power. Journal of Climate and Applied Meteorology 23(8), pp. 1184–1195.
15. Hill, D.C., McMillan, D., Bell, K.R.W., Infield, D. (2012). Application of auto-regressive models to U.K. wind speed data for power system impact studies. IEEE Trans Sustain Energy;3(1):134–41.
16. Tantareanu, C. (1992). Wind Prediction in Short Term: A first step for a better wind turbine control. Nordvestjysk Folkecenter for Vedvarende Energi, ISBN 87-7778-005-1.
17. Kavasseri, R.; Seetharaman, K. (2009). Day-ahead wind speed forecasting using f-ARIMA models. Renew. Energy, 34, 1388–1393.
18. Torres, J.L., A. Garcia, M. De Blas and A. De Francisco (2005). Forecast of hourly average wind speed with ARMA models in Navarre (Spain). Solar Energy 79(1), pp. 65–77.
19. Wu, Y.K., Lee, C.Y., Tsai, S.H. & Yu, S.N. (2010) Actual Experience on the Short-Term Wind Power Forecasting at Penghu-From an Island Perspective. Hangzhou, China: Proceedings of the 2010 International Conference on Power System Technology, 1–8.
20. Biermann, K., et al. "Entwicklung eines Rechenmodells zur Windleistungsprognose für das Gebiet des deutschen Verbundnetzes." Research Project by order of PTJ/BMU Germany (2005).
21. Zeng, J.W. & Qiao, W. (2011). Support Vector Machine-Based Short-Term Wind Power Forecasting. Phoenix, USA: Proceedings of the IEEE/PES Power Systems Conference and Exposition, , 1–8.

22. Xia, J.R., Zhao, P. & Dai, Y.P. (2010) Neuro-Fuzzy Networks for Short-Term Wind Power Forecasting. Hangzhou, China: Proceedings of the International Conference on Power System Technology, 1–5.
23. Pinson, P., Ranchin, T. & Kariniotakis, G. (2004). Short-term Wind Power Prediction for Offshore Wind Farms, , Chicago USA: Proc. of the 2004 Global Wind Power Conference.
24. ANEMOS (n.d.), ANEMOS Project Web Page Retrieved from http: //anemos.cma.fr/ Bacher, P., Madsen, H., Nielsen, H., (2009). Online short-term solar power forecasting. Sol. Energy 83, 1772–1783.
25. Pinson, P. & Madsen, H. (2009). Ensemble-based probabilistic forecasting at Horns Rev, Wind Energy special issue Offshore, Wind Energy, vol. 12, issue 2, pp. 137–155.
26. Kurt, M. (2017). Development of an Offshore Specific Wind Power Forecasting System, PhD Thesis, Kassel, Germany: Kassel University Press.
27. Dutton, J.A. (1976). The ceaseless wind: an introduction to the theory of atmospheric motion. New York: McGraw-Hill, , p 579.
28. Heinemann, D., Lorenz, E., & Girodo, M. (2006). Forecasting of solar radiation. In Solar Energy Resource Management for Electricity Generation from Local Level to Global Scale, pages 223– 233. Nova Science Publishers.
29. Chow, C.W., Urquhart, B., Lave. M. (2011). Intra-hour forecasting with a total sky imager at the UC3 San Diego solar energy testbed. Sol Energy, 85(11):2881–2893.
30. Hammer, A., Heinemann, D., & Hoyer, C. (2003). Solar energy assessment using remote sensing technologies. Remote Sens Environ 86:423–432.
31. Kühnert, J., Lorenz, E., & Heinemann, D. (2013) Satellite-based irradiance and power forecasting for the German energy market in solar energy forecasting and resource assessment.
32. Cali, Umit, and Vinayak Sharma. "Short-term wind power forecasting using long-short term memory based recurrent neural network model and variable selection." International Journal of Smart Grid and Clean Energy 8.2 (2019): 103–110.
33. Sarp, S., Kuzlu, M., Cali, U., Elma O., & Guler, O. (2021) "An Interpretable Solar Photovoltaic Power Generation Forecasting Approach Using An Explainable Artificial Intelligence Tool," 2021 IEEE Power & Energy Society Innovative Smart Grid Technologies Conference (ISGT), pp. 1–5, https://doi.org/10.1109/ISGT49243.2021.9372263.
34. Larson, D., Nonnenmacher, L., Coimbra, C.F.M., (2016). Day-ahead forecasting of solar power output from photovoltaic plants in the American Southwest. Renewable Energy 91, 11–20.
35. Lorenz, E., Heinemann, D., Wickramarathne, H., Beyer, H.G., Bofinger, S., (2007). Forecast of ensemble power production by grid-connected PV systems. Milano, Italy: 20th European PV Conference, 03.09.–07.09.2007.
36. Lorenz, E., Hurka, J., Karampela, G., Heinemann, D., Beyer, H.G., & Schneider, M., (2008). Qualified forecast of ensemble power production by spatially dispersed grid-connected PV systems. Valencia, Spain: 23rd European Photovoltaic Solar Energy Conference.
37. Lorenz, E., Scheidsteger, T., Hurka, J., Heinemann, D., & Kurz, C., (2011). Regional PV power prediction for improved grid integration. Prog. Photovoltaic.: Res. Appl. 19, 757–771.
38. Lorenz, E., Kühnert, J., Wolff, B., Hammer, A., Kramer, O., & Heinemann, D., (2014). PV power predictions on different spatial and temporal scales integrating PV measurements, satellite data and numerical weather predictions. Amsterdam, Netherlands: 29th EUPVSEC, 22.–26.
39. Dolara, A., Leva, S., & Manzolini, G., (2015b). Comparison of different physical models for PV power output prediction. Sol. Energy 119, 83–99.
40. Bacher, P., Madsen, H., Nielsen, H., (2009). Online short-term solar power forecasting. Sol. Energy 83, 1772–1783.
41. Chu, Y., Urquhart, B., Gohari, S., Pedro, H., Kleissl, J., Coimbra, C., (2015). Short-term reforecasting of power output from a 48 MWe solar PV plant. Sol. Energy 112, 68–77.
42. Pedro, H.T.C. & Coimbra, C.F.M., (2012). Assessment of forecasting techniques for solar power production with no exogenous inputs. Sol. Energy 86, 2017–2028

43. Boland, John and Korolkiewicz, Malgorzata and Agrawal, Manju and Huang, Jing (2012). Forecasting solar radiation on short time scales using a coupled autoregressive and dynamical system (CARDS) model. Australian Solar Energy Society

44. Bouzerdoum, M., Mellit, A., Massi Pavan, A., (2013). A hybrid model (SARIMA-SVM) for short-term power forecasting of a small-scale grid-connected photovoltaic plant. Sol. Energy 98, 226–235.

45. Bessa, R., Trindade, A., Silva, C., Miranda, V., 2015. Probabilistic solar power forecasting in smart grids using distributed information. Electrical Power Energy Systems, 72, 16–23.

46. Ying Y., & Dong, L. (2013). Short-term PV generation system direct power prediction model on wavelet neural network and weather type clustering, Hangzhou, China: IEEE International Conference on Intelligent Human-machine Systems and Cybernetics.

47. Yona, A., Senjyu, T., Saber, A.Y., Funabashi, T., Sekine, H., & Kim, C.H. (2007) Application of neural network to one-day-ahead 24 hours generating power forecasting for photo- voltaic system. IEEE Intelligent Systems Applications to Power Systems, 2007. ISAP 2007. International Conference on, pp. 1–6.

48. Oudjana, S.H., Hellal, A., & Mahamed, I.H. (2012). Short term photovoltaic power genera- tion forecasting using neural network. In: Environment and Electrical Engineering (EEEIC), 2012 11th International Conference on, pp. 706–711. IEEE.

49. Jang, R. J. (1991). Fuzzy Modeling Using Generalized Neural Networks and Kalman Filter Algorithm. Anaheim, CA, US: Proceedings of the 9th National Conference on Artificial Intelligence, 14–19. 2. pp. 762–767.

50. Shi, J., Lee, W.J., Liu, Y., Yang, Y., & Wang, P., (2011). Forecasting power output of photo- voltaic system based on weather classification and support vector machine. IEEE Industry Applications Society Annual Meeting (IAS).

51. Mohammed, A.A., Yaqub, W., Aung, Z., (2015). Probabilistic forecasting of solar power: an ensemble learning approach. Intelligent Decision Technol. – Smart Innovational Systems Technologies, 39, 449–458.

52. Wang, Jidong, et al. "Short-term photovoltaic power generation forecasting based on envi- ronmental factors and GA-SVM." Journal of Electrical Engineering and Technology 12.1 (2017): 64–71.

53. Zeng, J. & Qiao, W. (2013). Short-term solar power prediction using a support vector machine, Renew. Energy 52,118 e 127.

54. Predictive Maintenance: What is it & What are the Benefits?, https://www.onupkeep.com/learning/maintenance-types/predictive-maintenance.

55. Refaat, S.S. and Abu-Rub, H., 2016. Smart grid condition assessment: concepts, benefits, and developments. Power Electronics and Drives, 1(2), pp.147–163.

56. Refaat, S.S. and Abu-Rub, H., 2016. Smart grid condition assessment: concepts, benefits, and developments. Power Electronics and Drives, 1(2), pp.147–163.

57. Predictive Maintenance Technologies That Enhance Power Equipment Reliability, https://download.schneider-electric.com/files?p_enDocType=White+Paper&p_File_Name=998-2095-01-16-12AR0_EN.PDF&p_Doc_Ref=998-2095-01-16-12AR0_EN

58. Gohel, H.A., Upadhyay, H., Lagos, L., Cooper, K. and Sanzetenea, A., 2020. Predictive Maintenance Architecture Development for Nuclear Infrastructure using Machine Learning. Nuclear Engineering and Technology.

59. Hoffmann, M.W., Wildermuth, S., Gitzel, R., Boyaci, A., Gebhardt, J., Kaul, H., Amihai, I., Forg, B., Suriyah, M., Leibfried, T. and Stich, V., 2020. Integration of Novel Sensors and Machine Learning for Predictive Maintenance in Medium Voltage Switchgear to Enable the Energy and Mobility Revolutions. Sensors, 20(7), p.2099.

60. De Benedetti, M., Leonardi, F., Messina, F., Santoro, C. and Vasilakos, A., 2018. Anomaly detection and predictive maintenance for photovoltaic systems. Neurocomputing, 310, pp.59–68.

61. Bangalore, P. and Tjernberg, L.B., 2015. An artificial neural network approach for early fault detection of gearbox bearings. IEEE Transactions on Smart Grid, 6(2), pp.980–987.

62. Žarković, M. and Stojković, Z., 2019. Artificial intelligence SF6 circuit breaker health assessment. Electric Power Systems Research, 175, p.105912.
63. G. Balzer, "Condition assessment and reliability centered maintenance of high voltage equipment," Proceedings of 2005 International Symposium on Electrical Insulating Materials, 2005. (ISEIM 2005)., Kitakyushu, 2005, pp. 259–264 Vol. 1, doi: https://doi.org/10.1109/ISEIM.2005.193394.
64. Qingdong Feng and YongLiang Liang, "Condition assessment of substation equipment based on intelligence information fusion," IEEE PES Innovative Smart Grid Technologies, Tianjin, 2012, pp. 1–5, doi: https://doi.org/10.1109/ISGT-Asia.2012.6303093.
65. T. Huuhtanen and A. Jung, "Predictive Maintenance of Photovoltaic Panels via Deep Learning," 2018 IEEE Data Science Workshop (DSW), Lausanne, 2018, pp. 66–70, doi: https://doi.org/10.1109/DSW.2018.8439898.
66. A. K. Singh, Ibraheem, S. Khatoon, M. Muazzam and D. K. Chaturvedi, "Load forecasting techniques and methodologies: A review," 2012 2nd International Conference on Power, Control and Embedded Systems, Allahabad, 2012, pp. 1–10, doi: https://doi.org/10.1109/ICPCES.2012.6508132.
67. V. Dehalwar, A. Kalam, M. L. Kolhe and A. Zayegh, "Electricity load forecasting for Urban area using weather forecast information," 2016 IEEE International Conference on Power and Renewable Energy (ICPRE), Shanghai, 2016, pp. 355–359, doi: https://doi.org/10.1109/ICPRE.2016.7871231.
68. K. Zor, O. Timur and A. Teke, "A state-of-the-art review of artificial intelligence techniques for short-term electric load forecasting," 2017 6th International Youth Conference on Energy (IYCE), Budapest, 2017, pp. 1–7, doi: https://doi.org/10.1109/IYCE.2017.8003734.
69. C Deb, F Zhang, J Yang et al., "A review on time series fore-casting techniques for building energy consumption[J]", Renewable and Sustainable Energy Reviews, vol. 74, pp. 902–924, 2017.
70. R. J. Hyndman and G. Athanasopoulos, Forecasting: principles and practice. OTexts, 2018.
71. D.J. Sailor, "Relating residential and commercial sector electricity loads to climate evaluating state level sensitivities and vulnerabilities" Energy, 26 (2001), pp. 645–657.
72. B. U. Islam, "Comparison of conventional and modern load forecasting techniques based on artificial intelligence and expert systems," International Journal of Computer Science Issues (IJCSI), vol. 8, no. 5, p. 504, 2011.
73. H. Shi, M. Xu and R. Li, "Deep Learning for Household Load Forecasting—A Novel Pooling Deep RNN," in IEEE Transactions on Smart Grid, vol. 9, no. 5, pp. 5271–5280, Sept. 2018, doi: https://doi.org/10.1109/TSG.2017.2686012.
74. T. Khoa, L. Phuong, P. Binh, and N. T. Lien, "Application of wavelet and neural network to long-term load forecasting," in2004 International Conference on Power System Technology, 2004. Power Con 2004., vol. 1, pp. 840–844, IEEE,2004
75. W. Kong, Z. Y. Dong, Y. Jia, D. J. Hill, Y. Xu and Y. Zhang, "Short-Term Residential Load Forecasting Based on LSTM Recurrent Neural Network," in IEEE Transactions on Smart Grid, vol. 10, no. 1, pp. 841–851, Jan. 2019, doi: https://doi.org/10.1109/TSG.2017.2753802.
76. H. Shi, M. Xu and R. Li, "Deep Learning for Household Load Forecasting—A Novel Pooling Deep RNN," in IEEE Transactions on Smart Grid, vol. 9, no. 5, pp. 5271–5280, Sept. 2018, doi: https://doi.org/10.1109/TSG.2017.2686012.
77. K. Chen, K. Chen, Q. Wang, Z. He, J. Hu and J. He, "Short-Term Load Forecasting With Deep Residual Networks," in IEEE Transactions on Smart Grid, vol. 10, no. 4, pp. 3943–3952, July 2019, doi: https://doi.org/10.1109/TSG.2018.2844307.
78. Amarasinghe, K., Marino, D.L. and Manic, M., 2017, June. Deep neural networks for energy load forecasting. In 2017 IEEE 26th International Symposium on Industrial Electronics (ISIE) (pp. 1483–1488). IEEE.
79. Sun, M., Zhang, T., Wang, Y., Strbac, G. and Kang, C., 2019. Using Bayesian deep learning to capture uncertainty for residential net load forecasting. IEEE Transactions on Power Systems, 35(1), pp.188–201.

80. W. Kong et al., "Effect of automatic hyperparameter tuning for residential load forecasting via deep learning," 2017 Australasian Universities Power Engineering Conference (AUPEC), Melbourne, VIC, 2017, pp. 1–6, doi: https://doi.org/10.1109/AUPEC.2017.8282478.

81. C. Ye, Y. Ding, P. Wang and Z. Lin, "A Data-Driven Bottom-Up Approach for Spatial and Temporal Electric Load Forecasting," in IEEE Transactions on Power Systems, vol. 34, no. 3, pp. 1966–1979, May 2019, doi: https://doi.org/10.1109/TPWRS.2018.2889995.

82. C. Ye, Y. Ding, P. Wang and Z. Lin, "A Data-Driven Bottom-Up Approach for Spatial and Temporal Electric Load Forecasting," in IEEE Transactions on Power Systems, vol. 34, no. 3, pp. 1966–1979, May 2019, doi: https://doi.org/10.1109/TPWRS.2018.2889995.

83. D. Zhang, X. Han and C. Deng, "Review on the research and practice of deep learning and reinforcement learning in smart grids," in CSEE Journal of Power and Energy Systems, vol. 4, no. 3, pp. 362–370, September 2018, doi: https://doi.org/10.17775/CSEEJPES.2018.00520.

84. Junjian Qi, Shengwei Mei, and Feng Liu. "Blackout model considering slow process." In: IEEE Transactions on Power Systems 28.3 (2013), pp. 3274–3282.

85. T. Yu, B. Zhou, K. W. Chan, L. Chen and B. Yang, "Stochastic Optimal Relaxed Automatic Generation Control in Non-Markov Environment Based on Multi-Step Q(λ) Learning," in IEEE Transactions on Power Systems, vol. 26, no. 3, pp. 1272–1282, Aug. 2011, doi: https://doi.org/10.1109/TPWRS.2010.2102372.

86. Yu T, Xi L, Yang B, Xu Z, Jiang L. Multiagent stochastic dynamic game for smart generation control. J Energy Eng 2016;142(1), 04015012.

87. L. Xi et al., "A deep reinforcement learning algorithm for the power order optimization allocation of AGC in interconnected power grids," in CSEE Journal of Power and Energy Systems, vol. 6, no. 3, pp. 712–723, Sept. 2020, doi: https://doi.org/10.17775/CSEEJPES.2019.01840.

88. J. G. Vlachogiannis and N. D. Hatziargyriou, "Reinforcement learning for reactive power control," in IEEE Transactions on Power Systems, vol. 19, no. 3, pp. 1317–1325, Aug. 2004, doi: https://doi.org/10.1109/TPWRS.2004.831259.

89. Q. Yang, G. Wang, A. Sadeghi, G. B. Giannakis and J. Sun, "Two-Timescale Voltage Control in Distribution Grids Using Deep Reinforcement Learning," in IEEE Transactions on Smart Grid, vol. 11, no. 3, pp. 2313–2323, May 2020, doi: https://doi.org/10.1109/TSG.2019.2951769.

90. H. Xu, A. D. Domínguez-García and P. W. Sauer, "Optimal Tap Setting of Voltage Regulation Transformers Using Batch Reinforcement Learning," in IEEE Transactions on Power Systems, vol. 35, no. 3, pp. 1990–2001, May 2020, doi: https://doi.org/10.1109/TPWRS.2019.2948132.

91. J. Duan et al., "Deep-Reinforcement-Learning-Based Autonomous Voltage Control for Power Grid Operations," in IEEE Transactions on Power Systems, vol. 35, no. 1, pp. 814–817, Jan. 2020, doi: https://doi.org/10.1109/TPWRS.2019.2941134.

92. Y. Wang, "Grid Voltage Control Method Based on Generator Reactive Power Regulation Using Reinforcement Learning," 2020 IEEE/IAS Industrial and Commercial Power System Asia (I&CPS Asia), Weihai, China, 2020, pp. 1060–1065, doi: https://doi.org/10.1109/ICPSAsia48933.2020.9208556.

93. Ioannis Antonopoulos, Valentin Robu, Benoit Couraud, Desen Kirli, Sonam Norbu, Aristides Kiprakis, David Flynn, Sergio Elizondo-Gonzalez, Steve Wattam, "Artificial intelligence and machine learning approaches to energy demand-side response: A systematic review", Renewable and Sustainable Energy Reviews, Volume 130, 2020, 109899, ISSN 1364-0321.

94. D. Zhou, M. Balandat and C. Tomlin, "Residential demand response targeting using machine learning with observational data," 2016 IEEE 55th Conference on Decision and Control (CDC), Las Vegas, NV, 2016, pp. 6663–6668, doi: https://doi.org/10.1109/CDC.2016.7799295.

95. D. C. Mocanu, E. Mocanu, P. H. Nguyen, M. Gibescu and A. Liotta, "Big IoT data mining for real-time energy disaggregation in buildings," 2016 IEEE International Conference on Systems, Man, and Cybernetics (SMC), Budapest, 2016, pp. 003765–003769, doi: https://doi.org/10.1109/SMC.2016.7844820.

96. Heider Berlink and Anna Helena Reali Costa. 2015. Batch reinforcement learning for smart home energy management. In Proceedings of the 24th International Conference on Artificial Intelligence (IJCAI'15). AAAI Press, 2561–2567

97. D. O'Neill, M. Levorato, A. Goldsmith and U. Mitra, "Residential Demand Response Using Reinforcement Learning," 2010 First IEEE International Conference on Smart Grid Communications, Gaithersburg, MD, 2010, pp. 409–414, doi: https://doi.org/10.1109/SMARTGRID.2010.5622078.

98. Ozay, M., Esnaola, I., Vural, F.T.Y., Kulkarni, S.R. and Poor, H.V., 2015. Machine learning methods for attack detection in the smart grid. IEEE transactions on neural networks and learning systems, 27(8), pp.1773–1786.

99. Khan, S., Kifayat, K., Kashif Bashir, A., Gurtov, A. and Hassan, M., 2020. Intelligent intrusion detection system in smart grid using computational intelligence and machine learning. Transactions on Emerging Telecommunications Technologies, p.e4062.

100. Ford, V., Siraj, A. and Eberle, W., 2014, December. Smart grid energy fraud detection using artificial neural networks. In 2014 IEEE Symposium on Computational Intelligence Applications in Smart Grid (CIASG) (pp. 1–6). IEEE.

101. Esmalifalak, M., Liu, L., Nguyen, N., Zheng, R. and Han, Z., 2014. Detecting stealthy false data injection using machine learning in smart grid. IEEE Systems Journal, 11(3), pp.1644–1652.

102. Jindal, A., Dua, A., Kaur, K., Singh, M., Kumar, N. and Mishra, S., 2016. Decision tree and SVM-based data analytics for theft detection in smart grid. IEEE Transactions on Industrial Informatics, 12(3), pp.1005–1016.

103. Yan, J., Tang, B. and He, H., 2016, July. Detection of false data attacks in smart grid with supervised learning. In 2016 International Joint Conference on Neural Networks (IJCNN) (pp. 1395–1402). IEEE.

104. Kaygusuz, C., Babun, L., Aksu, H. and Uluagac, A.S., 2018, May. Detection of compromised smart grid devices with machine learning and convolution techniques. In 2018 IEEE International Conference on Communications (ICC) (pp. 1–6). IEEE.

105. Sakhnini, J., Karimipour, H. and Dehghantanha, A., 2019, August. Smart grid cyber attacks detection using supervised learning and heuristic feature selection. In 2019 IEEE 7th International Conference on Smart Energy Grid Engineering (SEGE) (pp. 108–112). IEEE.

106. He, Y., Mendis, G.J. and Wei, J., 2017. Real-time detection of false data injection attacks in smart grid: A deep learning-based intelligent mechanism. IEEE Transactions on Smart Grid, 8(5), pp.2505–2516.

107. Wei, J. and Mendis, G.J., 2016, April. A deep learning-based cyber-physical strategy to mitigate false data injection attack in smart grids. In 2016 Joint Workshop on Cyber-Physical Security and Resilience in Smart Grids (CPSR-SG) (pp. 1–6). IEEE.

108. DARPA1998 Dataset. 1998. Available online: http://www.ll.mit.edu/r-d/datasets/1998-darpa-intrusiondetection-evaluation-dataset (accessed on 16 October 2019)

109. KDD99 Dataset. 1999. Available online: http://kdd.ics.uci.edu/databases/kddcup99/kddcup99.html (Accessed 2019-10-16).

110. NSL-KDD99 Dataset. 2009. Available online: https://www.unb.ca/cic/datasets/nsl.html (Accessed 2019-10-16).

111. The UNSW-NB15 Dataset Description. 2018. Available online: https://www.unsw.adfa.edu.au/unsw-canberra-cyber/cybersecurity/ADFA-NB15-Datasets/ (Accessed 2019-10-16).

112. The ADFA Intrusion Detection Datasets. 2013. Available online: https://www.unsw.adfa.edu.au/unsw-canberra-cyber/cybersecurity/ADFA-IDS-Datasets/ (Accessed 2019-10-16).

Chapter 8
Foundations of Distributed Ledger Technology

In grid systems as designed in the mid-twentieth century, one party, the incumbent utility, was designated as the single repository for tracking energy production and consumption, for billing and load balancing purposes, and also over the long term for infrastructure planning. Smaller utilities realized an economic benefit from joining their generation and transmission assets, so they could focus strictly on distribution systems. The result was a split in the grid into the transmission system operator[1] (TSO) and the distribution system operator (DSO) or distribution utility, with generation provided by separate entities. This allowed each entity to focus on particular segments of the energy business, thereby achieving economic efficiencies. The generator/TSO/DSO system model has now become common in developed countries.

However, with the increasing availability of cheap renewable DERs such as solar panels, wind generators, and batteries, retail customers can now deploy their own carbon-free generation and storage equipment behind the meter, giving rise to the producer–consumer or prosumer. In many jurisdictions, the prosumer has two options: sell the power back to the utility or consume it on site. Recently, renewable energy and storage aggregators – third-party players who aggregate prosumer distributed energy resources and offer them to the distribution and transmission system operators – provide prosumers with additional options, for example, to bid their DERs into the ancillary services capacity market.

These developments have resulted in a distributed energy business ecosystem that is considerably more complicated and fragmented than when the original unified generation/transmission/distribution monopoly utility model was developed in the twentieth century. Energy systems today involve multiple parties providing value or consuming it in all parts of the business ecosystem, not simply a one-way flow of energy from the generator and distributor to the end customer, and a corresponding one-way flow of payments in the opposite direction back to the generator

---

[1] The transmission system operator (TSO) is known as the independent system operator (ISO) or regional system operator (RSO) in the USA.

U. Cali et al., *Digitalization of Power Markets and Systems Using Energy Informatics*, https://doi.org/10.1007/978-3-030-83301-5_8

169

and distributor. Keeping track of charges and energy flows using a centralized database is therefore becoming increasingly difficult.

An alternative that matches today's distributed energy business ecosystem more closely is the distributed ledger. In this chapter, we discuss the application of distributed ledger technology to energy systems. Section 8.1 presents the basics of distributed technology. Section 8.2 discusses the decentralized identifiers [1] and verifiable credentials [2], two new standards from the W3C standardization group for managing authentication and authorization in highly decentralized business ecosystems. Section 8.3 describes advancements in blockchain technology that have removed some of the limitations from the earlier years. Finally, Section 8.4 summarizes the chapter.

8.1   Fundamentals of Distributed Ledger Technology

A distributed ledger is a distributed database that is spread among multiple participants at different sites. The most well-known type of distributed ledger is the blockchain. In other words, blockchain technology (BT) can be considered as a subset of distributed ledger technology (DLT), as demonstrated in Fig. 8.1. A blockchain implements the distributed ledger database as a chain of information blocks linked together with cryptographic hashes. The hash links ensure that a block cannot be removed from the chain or another inserted before or after without the modification

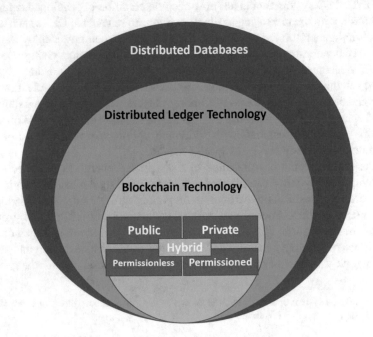

Fig. 8.1 Overview of distributed ledger and blockchain technology

being detected since otherwise, the hash link would differ. The result is a ledger that is tamper-evident and tamper-resistant. In this section, we present the fundamentals of distributed ledger technology.

Blockchain technology evolved out of the desire of some technologists in the financial community to decouple money from sponsorship by central governments. Precursors of blockchain technology were developed starting in the early 1990s, but blockchain first emerged into the international technology scene when the mysterious Satoshi Nakamoto released a whitepaper on Bitcoin in 2008. Bitcoin was the first cryptocurrency, a financial asset defined on a blockchain and secured by cryptography. The unit of cryptocurrency on the Bitcoin blockchain is called bitcoin. The first block for BTC, called the genesis block, was generated in 2009 and after one year, BTC was used to buy a pizza. IN 2011, BTC reached parity with the US dollar, and in 2012, the Bitcoin Foundation was launched to maintain and develop the blockchain technology underlying BTC.

BTC hit $1B market capitalization while In 2013, Vitali Buterin announced the Ethereum Project. The premise behind Ethereum was different than Bitcoin. The Ethereum blockchain was designed to be a "world computer", a decentralized computer on which anyone could write a program, called a smart contract, and execute code by paying a fee in cryptocurrency. Ethereum went public in 2014. Linux Foundation launched Hyperledger Fabric in 2015. The Decentralized Autonomous Organization (DAO) raised capital via crowdfunding in 2016. EOS launched while the Digital Trade Chain was announced by a consortium of European banks in 2017. Several businesses in various sectors started to implement blockchain use cases in 2018. The German government announced a national strategy document including a roadmap for a DLT centered digital transformation are communicated in 2019. Global COVD 19 pandemics prompted technologists to propose applications of blockchain to the health care sector (Fig. 8.2).

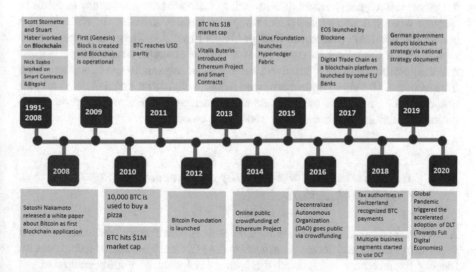

Fig. 8.2   Historical evolution of DLT and blockchain technology

8.1.1   What Are Blockchains Good For?

Blockchains originally rose to prominence through their deployment as the underlying distributed ledger used to record payment transactions for cryptocurrency systems. They have since been deployed to support a broad variety of other applications. However, not all business processes are candidates for deploying on a blockchain. A blockchain is a good choice for business processes with multiple parties involved that have three or more of the following properties [3]:

1. The parties require a common distributed database where information shared among the parties is written.
2. The parties have conflicting incentives or do not fully trust each other.
3. The lack of mutual trust requires a trusted third party (e.g., escrow service, data feed provider, licensing authority, notary public, etc.), which acts as a bottleneck and/or extracts fees.
4. The participants are governed by uniform rules.
5. Cryptography can be used for authenticating transaction participants and ensuring data integrity and non-repudiation.
6. The business process involves the transfer of a digital asset that would otherwise be difficult to track and measure.
7. An objective, immutable history or log of transactions is required.
8. Decision-making of the parties is transparent rather than confidential.
9. Oversight from a regulatory authority may be required.
10. Transaction rate does not exceed 5000 transactions per second.

In many cases, a blockchain applied to a business process replaces data entered into multiple, different databases, including Excel spreadsheets and email messages, maintained by multiple parties involved at different points in the lifecycle of the process. Such unorganized data recording methods have a tendency to result in disputes when one party loses data or happens to miss an email. In other cases, a centralized database maintained by a participant in the business ecosystem is the repository for the process's state, but if a dispute arises in which the participant maintaining the database is involved, the other participants in the ecosystem have little recourse. Having a single distributed ledger, with copies maintained by all parties, as the single "source of truth" about the business process, radically simplifies dispute resolution, consistency, and transaction processing.

8.1.2   Blockchain Platform Identity and Access Management (IAM) Architecture

An important design consideration for a blockchain architecture is whether and how to perform identity and access management (IAM). Consideration of techniques for identity management is required for both participants providing computational power and storage to maintain the blockchain and for participants conducting business on it. For those blockchain systems that require a participant to possess digital proof of identity, once a participant's identity has been authenticated, access control is required to limit the participant's access to only those operations authorized by their role.

IAM architectures for blockchain platforms can be classified according to the following:

- Permissionless/public: Participants don't need permission from any authority nor a verifiable offline identity to set up a node or conduct transactions. Transaction participants are identified by a naked public key not tied to any verifiable digital identity. Transactions are anonymous.
- Permissioned/public: These are of two types:

    - Anybody can set up a node, but only authorized participants can transact together. This is typically implemented by forking a separate distributed ledger side chain for the participants to transact.
    - Anybody can set up a node, but some nodes have special permissions. A network maintainer assigns specific permissions, unavailable to the others, to collections of identities, for example, to maintain the ledger state.
    - This is sometimes called a hybrid blockchain because it combines properties of both the permissonless/public and permissioned/private architectures.

In both cases, a verifiable digital identity is required.

- Permissioned/private: Only specific organizations are allowed to set up nodes, and only members of those organizations with verifiable digital identities are allowed to transact. Organizations and members have assigned roles, for example, a node may be permitted to maintain copies of the distributed ledger while an individual may be restricted to read-only access for auditing.

The fourth possible combination, permissionless/private, is somewhat of a contradiction in terms. "Permissionless" means that no identity is required to access and there are no access controls, while "private" means that access is restricted to only participants with a verifiable identity. While some blockchains claim to be permissionless/private, in fact, they are permissioned/public, as they allow anyone to access the platform (e.g., access is not restricted to members of specific organizations) but the participants need to have digital identities and can set up private groups for transacting among themselves, which are part of the ledger transaction rules.

There is widespread confusion around the difference between the privacy of transactions and access control in the blockchain world. "Permissioned" means identity management is required to access the ledger. Mechanisms to restrict transactions to certain parties, that is, to enforce privacy, such as forking a separate ledger, are part of the ledger transaction rules, and other techniques like encryption could be used, not part of the access control scheme. Similarly, "public" means access to the ledger is not restricted to any group of organizations, while "private" restricts access to certain organizations. While there are alternate classification schemes, the scheme presented here lines up more closely with conventional IAM terminology.

8.1.3 Examples of Widely Deployed Blockchain Platforms

The last few years have seen an explosion of blockchain technology platforms, many for cryptocurrency but also for other applications. Three platforms stand out as having the widest deployment: Bitcoin, Ethereum 1.0 and its permissioned block-chain forks, and Hyperledger Fabric. Future developments in blockchain technology may render one or the other of these platforms obsolete at some point, but for now, these platforms have the broadest adoption base.

Bitcoin was the first blockchain platform to achieve widespread deployment and is perhaps the best known [4]. It went live in 2009 and is only used for cryptocurrency transactions. Transactions are programmed in a simple Fourth-like language called Script. Anyone can bring up a validation node or validator, and anyone can set up an account on the Bitcoin blockchain without presenting off-chain proof of identity, so the access control architecture is permissionless/public. Transaction processing operates as follows. Transactions are grouped into blocks at intervals of around 10 min. A validator validates a transaction then broadcasts the transaction to the rest of the nodes in the network for consensus. Agreement among nodes about what transactions to enter into the distributed ledger is achieved using a distributed consensus algorithm called proof of work (PoW). The validators are sometimes called miners because the validator that successfully completes the PoW for a transaction is rewarded with some bitcoin cryptocurrency. The PoW algorithm is discussed in Section 8.1.4.3.2.

Ethereum 1.0 is the next most widely adopted blockchain platform [5], and it also supports a permissionless/public access control architecture. Unlike Bitcoin, the Ethereum blockchain was designed from the beginning to be a decentralized ledger for applications other than cryptocurrency, and it supports a decentralized execution environment based on the Ethereum Virtual Machine (EVM). The EVM has a standard language-level virtual machine architecture similar to the Java Virtual Machine (JVM) for Java. Bytecodes are sequentially executed, with each block-chain node executing the same bytecode in step. This ensures that the state update is sequential and deterministic but imposes an inherent performance bottleneck. Programmers use a high-level, Turing-complete language called Solidity to program the EVM with smart contracts that govern interactions between the participants. While Ethereum's overall architecture is permissionless, smart contracts can impose fine-grained access control on the participants if desired. Ethereum supports two kinds of accounts: an account for an external participant like a person or company and an account for a smart contract. Ethereum has its own cryptocurrency, the ether, but ether is primarily for controlling execution time on the decentralized computation fabric since, without any limit, a program in an infinite loop could lock up the entire network. Ethereum 1.0 nodes also use PoW for consensus.

Ethereum has spawned a variety of spin-off projects with different access control models. Quorum [6] was originally developed by J.P. Morgan as a fork of Ethereum but was open-sourced and is now supported by a number of groups. Quorum has a permissioned/private access control architecture and requires transaction nodes to authenticate to join the network and participants to set up an account linked to an off-chain identity. One of the main consequences of more restrictive access control is that distributed consensus algorithms having better performance can be used [7], and Quorum allows a variety of distributed consensus algorithmic plug-ins. Another consequence is that Quorum also supports private transactions, which are necessary for some business-related applications [8]. In addition, Quorum does not support the ether cryptocurrency since limiting computation time is unnecessary. Quorum has been widely used by companies in the financial industry for prototyping systems.

Because Hyperledger Fabric was designed from the beginning on a permissioned/private access control model, its overall architecture is quite different from the other blockchain platforms [9]. Fabric was developed by IBM, open-sourced through the Linux Foundation's Hyperledger blockchain project, and does not support a cryptocurrency. The Fabric design is more like an operating system, where developers can program applications in general-purpose languages like Java or Python, but Fabric also supports Solidity. Fabric applications access the blockchain through software development kits (SDKs) in the respective language, and its unique transaction processing architecture is discussed in detail in Section 8.1.4. Fabric is widely deployed in situations like supply chain management, where a business process involves multiple organizations. The organizations typically form a consortium to govern blockchain deployment and access.

Blockchain platforms have a not undeserved reputation for low performance. Table 8.1 provides comparative performance numbers for the blockchain platforms discussed above compared with the commercial Visa credit card transaction processing system. As the table indicates, the blockchain systems are roughly an order of magnitude, or more, slower than Visa. For the permissioned, private blockchains, some performance optimizations are possible through specific deployment architectures. However, blockchain technology in general and performance, in particular, is under active and intensive research, and recently some research blockchain platforms have announced considerably improved performance. The Tectum platform, for example, advertises performance in excess of 1 million transactions per section [10].

Table 8.1 Comparative transaction processing performance

Transaction processor	Maximum transactions/second
Bitcoin	7
Ethereum 1.0	30
Quorum	2300
Hyperledger fabric	20,000
Visa	65,000

8.1.4   Basic Blockchain Mechanisms

Underlying a blockchain platform are three foundational mechanisms:

- A distributed ledger that contains the state agreed upon by all nodes and that is connected into one global, tamper-resistant data structure,
- Cryptography for verifying transactions, enabling auditing and traceability, and for other purposes such as connecting the data blocks into a chain,
- Agreement between all parties on what state is committed to the ledger through a process of distributed consensus.

Once the state is committed to the blockchain, it can never be modified and is readable by all parties. In the following subsections, we discuss these mechanisms in more detail.

8.1.4.1   Blockchain Representation

A high-level architecture of a blockchain distributed ledger data structure is shown in Fig. 8.1. The data structure consists of a sequence of blocks linearly growing in time. Each block has two parts: the block header containing data controlling the chain and the block data containing data committed by participants in transactions entered during a particular time period.

The block header contains two hashes:

- A hash of the previous block header that serves as a link between the current block and the previous block,
- And a hash of the block data, which ensures that the block data remains tamper-proof.

The timestamp indicates when the block was committed to the chain, and timestamps in sequential blocks are increasing. Transactions in the block data section were entered in the time period between the previous block header timestamp and the timestamp in the current block header.

An additional section of control data provides support for different types of consensus algorithms and other details of the blockchain control design.

The actual implementation of this architecture for a particular blockchain platform depends on the specific data structures and algorithms chosen by the platform designers, just as the architecture of a relational database is a table with schema items in the columns and data entries in the rows while the implementation is based on data structures such trees and their respective algorithms (Fig. 8.3).

8.1.4.2   Uses of Cryptography

Blockchain systems use cryptography to perform a number of functions. Cryptographic hashes implement the following functions:

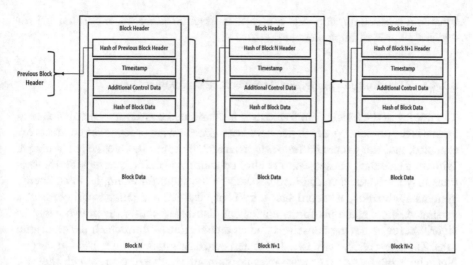

Fig. 8.3 Architecture of a blockchain data structure

- Hashes are used as links between blocks in the block headers to ensure that an existing block can't be deleted nor a new one inserted between two already linked blocks.
- A hash of the block data in the block header protects data committed to the chain from tampering.
- In blockchains that use proof of work (PoW) for their distributed consensus algorithm, cryptographic hashing plays a central role, as discussed in Section 8.1.4.3.2.

The Ethereum blockchain uses the elliptic curve Keccak-256 algorithm, similar to but not the same as the SHA-3 algorithm that is specified in the FIPS-202 NIST standard [11] for these purposes. Some blockchain systems calculate the address of a participant's account by hashing the participant's public key, for example, Ethereum [12].

Public key cryptosystems are used for signing and authenticating transactions in blockchain platforms. Transactions are signed with the transaction originator using their account's private key and verified with their account's public key. Having a cryptographically verifiable signature enables auditing, regulatory examination, and dispute resolution. Most blockchain systems today use the elliptic curve digital signature algorithm (ECDSA) with 256-bit keys (ECC-256) for their cryptosystem, though they differ in the elliptic curves and parameters employed [13]. Transaction data is also encrypted in transit, but blockchain platforms do not, by default, provide confidentiality of transaction data at rest. Some platforms support additional services that do allow confidentiality of transactions recorded on the chain. For example, Hyperledger Fabric has channels that allow only certain, authenticated parties to participate [14], and Quorum, a permissioned version of the permissionless

Ethereum blockchain, supports a private transaction processor algorithm [8] for which a number of implementations exist.

8.1.4.3 Distributed Consensus Architecture and Algorithms

The last leg in the blockchain foundation is distributed consensus. Distributed consensus is the process by which the validators agree in what order smart contracts are executed, and the generated transactions are written into the blockchain's ledger. Without distributed consensus to control concurrent updates, overlapping transactions may introduce a causal inconsistency in the state; for example, in an energy systems application, a record that a kWh of electricity is purchased before it is generated. Blockchain platforms implement distributed consensus in two parts: an algorithm for achieving consensus and an architecture within which the algorithm runs. There are really only two distributed consensus architectures used in current blockchain platforms: the order–execute–commit architecture on which the vast majority of blockchain platforms are built and the execute–order–validate–commit architecture, which is exclusively used by Hyperledger Fabric. These correspond to the pessimistic and optimistic concurrency architectures in traditional database systems, respectively. On the other hand, quite a few distributed consensus algorithms have come out of the distributed systems research community over the years, and many of these have been deployed in blockchain platforms. In the next two subsections, we drill down on these two topics in more detail.

Distributed Consensus Architectures

The first key part of the blockchain distributed consensus design is the distributed consensus architecture. Most blockchains implement an order-execute-commit architecture that consists of the following steps:

1. Requests to execute a smart contract or write some data to the distributed ledger state are broadcast to all validators in the blockchain during the current transaction time period.
2. At the end of the transaction time period, the nodes run their distributed consensus algorithm and generate a causal order in which to execute the transactions.
3. All the nodes execute the smart contracts in the designated order at the same time, and the results are committed to the distributed ledger state.

This architecture is one of the main reasons why performance in currently deployed blockchain platforms lags behind that of traditional transaction processors. In a traditional transaction processor, there is a single logical database (which may be physically replicated for reliability and failover) that performs the transaction. In a blockchain system with thousands of validators, each node needs to wait until all the other nodes are finished ordering the transaction and executing it before the transaction can be declared committed and written to the blockchain's state. Every node deterministically executes the same smart contracts and commits the same state in exactly the same order in the same time interval to ensure the distributed ledger state is consistent with a causal ordering. Communication delays between the validators impose an additional delay on time required to complete a transaction.

In contrast, Hyperledger Fabric is built on an execute–order–validate–commit architecture [9]. In Hyperledger Fabric, smart contracts, which Fabric calls chaincode, are first required to undergo an endorsement process before being submitted for ordering and validation. Subsets of transaction endorsers simulate the transactions for clients using copies of the main chain state. The simulation results in a set of read and write operations to the chain state, ordered in time. When the endorser completes the simulation, it returns a cryptographically signed endorsement to the client with the read and write sets. The client then submits the transaction together with the collected endorsements to the ordering service. The number of endorsements that a client must collect before submitting the transaction to the ordering service is governed by the endorsement policy for the particular Fabric deployment.

The ordering service establishes a total order over all transactions submitted during a particular period through distributed consensus among all ordering nodes. Transactions are then batched into blocks and submitted to the validation service. The validation service performs three operations on the transaction: checks for adherence to the endorsement policy, checks to ensure that no read–write conflicts occur, that is, that the transactions in the block are causally consistent, and then commits the block to the distributed ledger. By separating out the distributed consensus from transaction processing, Fabric's execute–order–validate–commit architecture can achieve more parallelism than platforms that are built on an order–execute–commit architecture.

Distributed Consensus Algorithms

Distributed consensus algorithms are the second key part of the blockchain consensus design. The distributed consensus algorithm is the mechanism by which the blockchain validators decide what transactions to accept and how to impose a causal ordering on read and write operations. Distributed consensus is especially difficult to enforce for public blockchain platforms like Bitcoin and Ethereum since the platform can't assume that all validators are cooperative. Because anyone can set up a server that joins the network as a validator, it is possible that someone might join the network with one or several nodes for the sole purpose of disrupting the consensus process. Public blockchain platforms, therefore, have a very limited choice of distributed consensus algorithms and are mainly based on two: proof of work (PoW) or proof of stake (PoS). As shown in Fig. 8.4, distributed consensus algorithms differ depending on the authentication policy of the blockchain platform.

PoW algorithms require validators or miners (as validators are called in cryptocurrency blockchains) to present proof that they have accomplished a difficult computational task. The Bitcoin PoW algorithm requires the miner to take the SHA-256 cryptographic hash of a block of transactions repeatedly until the first n bits are all

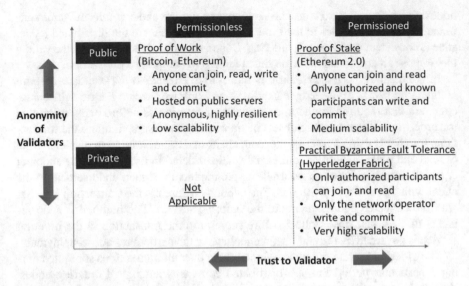

Fig. 8.4 Overview of DLT types

zero, with n increasing at periodic, longer-term intervals to increase the difficulty of the task. The first miner to complete this process submits the PoW along with the block to the other miners for incorporation into the distributed ledger and receives some bitcoin as a reward. The Ethereum PoW algorithm, called Ethash, is somewhat more complicated and requires miners to randomly grab chunks of a dataset generated from a smaller pseudorandom cache and hash the chunks together [15]. While the Bitcoin PoW algorithm depends on raw computational power to solve a very narrowly defined problem, the Ethereum algorithm depends more on how quickly and efficiently the hardware on which the validator is running can move data around in memory.

PoW algorithms have been criticized for being a massive waste of electricity. For example, in fall 2020, the amount of electricity for mining Bitcoin was estimated at 33.44 TWh annually [16]. Another problem with PoW algorithms is that they reward organizations that can deploy massive amounts of special-purpose hardware with chips specifically designed to solve the computational problem, cutting out smaller organizations and diverting engineering and production resources into economically useless projects. A modification of the PoW algorithm that doesn't suffer from these drawbacks is proof of useful work. The proof of useful work algorithm requires a miner to prove that it has performed some work that solves a problem submitted by a client [17].

Machine learning algorithms require lots of computation and therefore are good candidates for proof of useful work algorithms. The proof of useful work algorithm also requires some special-purpose hardware, namely processor support for secure enclave computation, so that attackers can't disrupt the computation. However, unlike the hashing ASICs developed specifically for bitcoin mining, secure enclave hardware is used for many other purposes and is available on widely deployed processor architectures such as the Intel architecture [18]. However, no widely deployed public blockchain has yet incorporated proof of useful work into its platform.

The PoS algorithm is another solution to the wasteful nature of PoW. PoS has been adopted by Ethereum 2.0 [19]. Some other permissionless blockchain systems also use PoS. In PoS consensus, if a node wants to become a validator, it must lock up a certain amount of cryptocurrency – its stake – in an escrow account. If the block is accepted, the validator collects a transaction fee. If the validator tries to disrupt the protocol, however, its stake is confiscated, and it is banned from the blockchain.

Proof of Authority (PoA) is a consensus mechanism related to PoS. Instead of staking cryptocurrency, the validators stake their reputations. Validators are selected based on their reputation, and the pool of validators is limited in size. A node is admitted to the validator pool by passing a rigorous test that is standardized and sufficiently difficult to pass that it can weed out questionable validators to ensure long term commitment. The advantage of PoA over PoS is that with PoS, nodes don't have to reveal the ratio of their stake to their total holdings, so there is no information about whether they might actually willingly forfeit their stake because it is such a small part of their total holdings. PoS can be used in either public or private blockchains.

In contrast to public blockchains, private blockchains can use much more efficient algorithms for consensus. Because the validators are vetted and known to be trustworthy, other non-validator participants can depend on the validators without requiring the validators to perform energy-consuming PoW or requiring the validator to deposit a stake for PoS. Most private and permissioned blockchains use some variation of partial Byzantine consensus (also known as partial Byzantine fault tolerance, PBFT) [20]. Partial Byzantine consensus is a process by which a collection of nodes in a distributed system can come to an agreement even though some nodes can fail in arbitrary and malicious ways. Many partial Byzantine consensus algorithms depend on a leader election, whereby one or several validators are periodically elected to perform the transaction validation, and the leaders are rotated periodically.

Starting the late 1990s, the distributed systems research community has developing a collection of algorithms that fulfill some or all of the characteristics necessary for achieving Byzantine consensus, and distributed consensus for permissioned/private blockchains is an area of active research which is likely to see continuing improvement in the future.

8.1.5   *Smart Contracts*

While a distributed ledger for a passive recording of data has numerous benefits, the ability to do distributed computation on the ledger state within the context of the blockchain can provide vastly more functionality, particularly in cases where two or more parties have contractual obligations based on changes in the ledger state. Smart contracts were added to blockchains to provide such an executable extension. A smart contract is a computer protocol intended to digitally facilitate, verify, or enforce the negotiation or performance of a contract. Smart contracts correct a flaw in the original Internet architecture: the lack of any protocol or mechanism to account for decentralized value creation and transfer.

However, though smart contracts were originally designed for the narrow purpose of digitally enforcing contractual obligations, their use has expanded as blockchains supporting smart contract languages have been applied to a broader variety of applications. A smart contract really functions as a general-purpose distributed computation mechanism that modifies state on a blockchain. The introduction of smart contracts transformed the blockchain from a distributed database into a platform for decentralized applications, much as in the early 2000s the introduction of asynchronous JavaScript, and XML (AJAX) turned the World Wide Web from an archive of passively hyperlinked pages into a platform for Web applications.

Blockchain systems support smart contracts in a variety of ways. One common way is to provide a simple stack-based language implemented as a virtual machine or interpreter. The Bitcoin Script language and the Ethereum Virtual Machine (EVM) for Ethereum 1.0 are examples. The EVM is a language virtual machine similar to the Java Virtual Machine, with a Last In-First Out (LIFO) stack having 1024 elements and 256-bit bytecodes. The Bitcoin opcode set is not Turing complete, while the EVM bytecode set includes codes for loops. As a consequence, smart contracts in Script, the Bitcoin language, are limited in functionality. For example, you can program a transaction for Bitcoin transfer to not complete until more than one participant has signed off on it. But programming in Script is difficult, similar to programming in an assembly language. In addition, it has little support for security, so it has seen very little use in deployment. In 2019, a language called Miniscript was developed as a higher-level language that compiles down into Script and includes security guarantees and optimization of operations for performance. Since the appearance of Miniscript, other higher-level languages have been developed that compile to Miniscript [21].

The most popular domain-specific smart contract language is Solidity [22], which was introduced with the Ethereum blockchain. Solidity is a high-level, object-oriented language syntactically similar to Java or JavaScript. Solidity compiles into bytecodes for the EVM, is statically typed and features classes, multiple inheritances, method scoping (public, private, etc.), and methods that run before and after the main method, all similar to many general-purpose, object-oriented languages.

However, it has some peculiarities as a result of executing on the blockchain. An additional *Transaction* method scope only allows the method to execute within a transaction, and Solidity methods have no access to any i/o except to and from blockchain state. Any state can only be passed in parameters to the initial method call or read from and written to the blockchain. In addition, Solidity programs have no main function or method where execution starts. Execution must be kicked off from code written in a general-purpose language like Javascript, then a remote procedure call (RPC) made to the blockchain to start the Solidity code. A Solidity execution thread can be terminated early by invoking a method on an object of built-in-type *Event*. The *Event* terminates the running transaction and throws the execution thread back into the RPC handler from which the original call to Solidity started, similar to an exception in C++ or Java.

Hyperledger Fabric and a few other blockchain platforms allow smart contracts to be written in a general-purpose language, like C++, Javascript, or Python, and run using the normal language runtime. Fabric provides SDKs providing blockchain operations and access for several general-purpose, high-level languages, and programmers can develop applications in the language they feel most comfortable programming. The SDK takes care of authentication and authorization required by the Fabric access control architecture and supports operations to act as a node in the Fabric blockchain or to submit transactions. Allowing programmers to use general-purpose languages is consistent with Fabric's design goal of supporting an operating system-like environment for programming applications on the blockchain.

With Ethereum 2.0, the same support for common programming languages is available through the use of Ethereum Web Assembly (Ewasm) [24], a version of WebAssembly [25] with features to support smart contract programming on the Ethereum 2.0 blockchain. WebAssembly has the same stack-based language virtual machine implementation with sandboxing as the EVM, but has a much wider support base, as it was designed for running in browsers. All major browsers support it, and while it was originally developed to support Javascript, compilers in various stages of development now exist for over 30 languages, including C, C++, Go, Rust, and Python, in addition to Solidity. Ewasm modifies WebAssembly by removing floating point support and any constructs that could lead to nondeterminism and by adding instructions for metering and for interface to the Ethereum blockchain environment. In addition, Ewasm specifies semantics for an *ewasm* contract, a system contract, and backward compatibility with EVM. The metering support is required in order for Ethereum to charge for transactions based on the number and type of instructions in the contract.

Unlike Web applications and enterprise client/server applications, applications written on a blockchain computation platform are completely decentralized and therefore are called decentralized applications, or *dapps*. From a logical standpoint, no centralized server, even replicated, hosts a dapp. From a deployment perspective, of course, the code for the dapp runs on a server, but no visible distinction is ever exposed between node state on different servers. Figure 8.2 illustrates the three-tier architecture of a Solidity dapp running on either the Ethereum 1.0 or Quorum blockchains:

- In the top layer, standard application code written in a high-level language that may run in a Web browser, handles the UI, communication, and non-blockchain i/o.
- An RPC mechanism between the standard application code and the Solidity code transfers control to and from the blockchain. The RPC mechanisms shown in the figure include JSON RPC [26] and representational state transfer (REST) HTTP-

based RPC [23]. Alternatively, the RPC may be local to a blockchain gateway running on the local server. For Solidity, the Web 3.0 SDK is often used for conducting the RPC [27, 28].
• On the bottom tier is the smart contract code running in the EVM, and changes to the blockchain state are made through the storage manager.

Dapps constitute a whole new class of applications, different from enterprise client/server and Web applications, based on a decentralized execution architecture.

8.1.6 Tokenization

Business processes that deal with the transfer of digital assets are well suited for blockchain implementation. A digital asset is an asset defined by a finite set of quantifiable properties that primarily exists in digital form, though it may be tied back to some actual asset or process that is nondigital. For example, renewable energy credits (RECs) are digital assets that are created by measuring the generation of renewable, carbon-free electricity from a renewable energy provider and offered for sale to an energy provider that needs to offset generation from fossil fuels to comply with regulatory requirements. A blockchain is an excellent way to track such assets because it provides accountability and transparency, including to the regulatory authorities.

While there are various ad hoc ways to track digital assets on a blockchain platform, tokenization has become increasingly popular. The record of the digital asset on the blockchain is called a token, and a token is often implemented as state associated with an account owned by a participant. Accessing the token is only possible through a smart contract, and smart contracts are used to transfer the token from one participant to another. Some blockchain platforms support tokenization for specialized purposes, but the Ethereum 1.0 platform has the most general-purpose tokenization support. Ethereum 1.0 supports two basic classes of tokens, fungible tokens, and nonfungible tokens.

For fungible tokens, each token has exactly the same value and a token can be subdivided up to some minimum unit. The Ethereum Request for Comments-20 (ERC-20) standard [29], as amended and updated by later standards, describes how to create and manage fungible tokens on the Ethereum 1.0 platform. An example of a fungible token is a dollar bill. One dollar bill is worth exactly the same as another, and a dollar can be subdivided up to a minimum unit of one cent. Another example is the Ethereum native cryptocurrency, the ether. Ethereum also supports many other kinds of cryptocurrency based on fungible tokens. In contrast, nonfungible tokens each have a unique value that is different from the others. Examples of nonfungible tokens are tokens that track physical property, such as land, or debts, such as or tokens that are tied to unique works of art. The standard for Ethereum nonfungible tokens is ERC-721 [30]. Finally, some digital assets have both a fungible and nonfungible part. Rather than representing the asset as two separate tokens, the

ERC-1155 [31] multi-token standard specifies a protocol for bundling the fungible and nonfungible parts so they can be transferred together.

8.2   Decentralized Identifiers and Verifiable Credentials

Chapter 5 discusses identity and access management (IAM) for smart grid systems where the smart grid system is run by a single organization and a single identity database managed by the organization is sufficient for holding the authentication credentials and authorization permissions for the principals. Single organization IAM systems can be extended to a small number of suppliers or customers through service accounts and public key certificates if necessary, but scaling up such systems to decentralized business ecosystems where hundreds of organizations participate can become a challenge.

Decentralized identifiers [1] offer a promising solution to the problem. A decentralized identity service deployment is highly scalable, so it should not depend on the number of participants in the business ecosystem. Verifiable credentials [2] fulfill the same function for authorization. A verifiable credential makes a digitally verifiable statement about the properties of a decentralized identity. This statement can be used for various authorization and access control purposes. While both decentralized identifiers and verifiable credentials are quite new, they hold promise for the evolving the decentralized business ecosystem around renewable energy resources and decarbonized electrical grids. This section provides an introduction to decentralized identifiers and verifiable credentials.

8.2.1   Decentralized Identifiers (DIDs)

A decentralized identifier (DID) is an identifier that is independent of any organization's (even a federated identity provider's or PKI's) siloed identity management system. It can identify a person, organization, hardware device, software package, or, in principle, anything a principal setting up the identifier wants to identify. A DID is a uniform resource locator (URL) format object having the scheme designator . *For example*:

did:*example:0123456789abcdef*

The *example* field is the DID method name. The DID method contains algorithms for forming the DID, verifying it, and looking up a DID document where more details on the subject can be found. The long sequence of letters and numbers is the actual identifier itself and is generated in a way that makes it universally unique to a high probability, that is, it is a statistically unique identifier.

The DID is bound to a DID document containing the details of the identity to which the DID belongs, known as the DID's subject. The DID document is structured as a JSON-LD [32] document and includes only one required field containing

DID to which the document is bound. It can also contain collection of the following possible optional fields:

- An array of records containing public keys used for authenticating and authorizing updates to the DID document or establishing secure communication from the subject to service endpoints,
- An array of authentication methods to use when communicating with the DID subject,
- A collection of proof methods to establish the authenticity of the DID document itself,
- A list of service endpoints (URLs) that can represent any type of service the subject wants, for example, a cryptocurrency blockchain URL and account address for transferring payment.

DID documents are kept in document repositories. When a client wants to resolve a DID to its document, it queries a DID resolver, similar to how DNS names are resolved to IP addresses. The DID resolver consults the DID method service to provide the document's location and returns the document. A universal resolver contains method resolvers for all known methods and resides at a well-known DNS name and IP address. New DID methods can be defined according to the W3C standard [33], and as of fall 2020, there were a total of 47 active provisional DID methods registered. Most of the DID methods are implemented on a blockchain platform, utilizing smart contracts for implementing parts of the DID method and the distributed ledger for storing mappings to DID documents, but the DID standard has no requirement for a DID method to be implemented on a blockchain.

DIDs are similar to public key certificates, with a few notable exceptions. With public key certificates, trust is based on verification of the certificate chain and the trustworthiness of the CAs in the chain. With DIDs, trust is based on verification of the signature on the DID and on the trustworthiness of the DID method network. The subject controls the DID through their private key, and the DID can only be modified by the subject. In a PKI, the subject's private key only binds the subject to the CA for purposes of identity verification. The subject can't change its certificate in most cases, as that is typically under the control of the CA that issued the certificate. Finally, DIDs have no certificate revocation chains and no expiration date. While this might seem to open DIDs to a larger probability of attack if the subject loses control of the key, it actually provides DIDs with more security since a CA can't arbitrarily revoke a subject's DID nor is it possible for someone to forget to renew the certificate, causing systems to fail identity checks. A PKI subject can lose control of its key too, and unless the compromise is reported to the CA, the imposter can pose as the subject until the CA finally revokes the certificate. With a DID, the subject can revoke its DID as soon as it discovers the compromise by simply contacting its DID method resolver.

8.2.2 Verifiable Credentials

Physical credentials like passports and drivers' licenses are widely used to assert claims about the principal possessing the credential. For example, a passport asserts a claim about a person's nationality, which then grants them the right to work in a particular country and the right to travel between countries. Verifiable credentials extend the ability to assert such claims to digital documents in a way that allows cryptographic verification of the claim and preserves the privacy of the subject by only exposing information related to the claim. Verifiable credential documents are defined in the JSON-LD format similar to DID documents.

A verifiable credential document contains a field defining the credential's subject, which is usually a DID URL containing the subject's DID. The document also includes a URL with a link to the issuer and a list of claims about the subject. For example, if the credential claims the subject has a Bachelor of Science degree, the URL may include a link to the Records Department of the university that issued the degree, and more detail on the degree claim, like the major and minor subject areas, when the degree was granted, etc.

Verifiable credentials can be stored anywhere, but one convenient place is in identity hubs [32]. An identity hub is an accessible store for safely and securely storing objects associated with the subject's identity and claims, like DID documents and verifiable credentials. A DID subject indicates that its claims are stored in an identity hub by including an identity hub URL in its list of service endpoints. When the DID subject requests authorization of a claim, the authorizing entity resolves the DID to the DID document, then accesses the identity hub link in the service endpoints list to obtain the verifiable credential document associated with the claim.

8.3 Improvements in Blockchain Technology

8.3.1 Forking

Blockchain technology continues to improve as researchers continue to make progress in solving outstanding problems. Since blockchains are a very recent technological innovation, in comparison with, say, the Internet, large public blockchains such as Ethereum are likely to see major changes, and, in fact, Ethererum underwent a major technological change from version 1.0 to version 2.0 in 2020. However, unlike information technology systems owned by large Internet companies with a huge public audience, Ethereum in some sense represents a public utility. Any changes in the platform are likely to be opposed by frequent users, especially if the distributed ledger format or schema are changed. Bitcoin, for this reason, has resisted any large changes in its technology base, though there were a few in the early years to fix bugs in the protocol and code.

A change in the blockchain code that causes nodes running the old version to process distributed ledger entries differently from a node running the new version is called a *fork*. When a version change in the blockchain code causes nodes running the new version to enforce additional rules that the nodes running the old version don't, the fork is termed a *soft fork*. With a soft fork, the version change is backward compatible in the sense that nodes running the old code can continue to process blocks. Soft forks allow nodes to incrementally upgrade. If, however, the code change causes the nodes running the old code to no longer be able to process distributed ledger entries, the fork is a *hard fork*. A hard fork requires all the nodes to be upgraded at once; otherwise, the blockchain will effectively be split into two.

There have been a number of soft and hard forks in the history of Bitcoin and Ethereum. In the case of Bitcoin, many of the forks were intentional and designed to offer a completely new cryptocurrency with new features, called an *altcoin*. For example, Bitcoin Cash is a successful fork of Bitcoin in August 2017 that was intentional [35]. A hard fork of Ethereum occurred in response to the DAO hack. An organization called The DAO had raised over $100 million in cryptocurrency, part of which was stored as ether (ETH) on the Ethereum blockchain. Due to bugs involving recursive calls to smart contracts, $50 million was stolen from the DAO account on June 17, 2016. The Ethereum Foundation fixed the bug, but it required a rewriting of the blockchain's distributed ledger, which is supposed to be immutable. A subgroup of the Ethereum community opposed the rewriting and instituted a hard fork, forming the Ethereum Classic blockchain and a new cryptocurrency type, the ETC.

Minor forks can occur if two miners successfully validate a block and submit it for acceptance at the same time. Some groups of nodes may receive the block from one miner and accept it, while the other group may receive and accept the block from the other miner. Communicating between nodes takes a finite amount of time, so nodes that are closer to one miner are more likely to accept the block they receive first. Such minor forks happen all the time and ultimately die out because the majority of nodes decide to accept one chain or the other, usually the longest chain. The blocks on the fork that are not accepted are called stale blocks. The transactions in stale blocks are returned to the pending transaction list and are incorporated into valid blocks on the main chain.

8.3.2 Performance Improvements in Ethereum 2.0

As discussed briefly in Section 8.1.3, the first generation of blockchain platforms have a deserved reputation for poor performance. While commercial credit card processors such as Visa have a theoretical peak performance of 65,000 transactions per second (tps) [36], and an average performance of around 1700 tps, Bitcoin only manages to achieve 4.5 tps on average with a maximum rate of 7 tps. Performance is especially a problem with permissionless/public blockchains. Hyperledger Fabric has been shown to scale to 20,000 tps maximum [37] with an average of around

2400 tps. The relatively poor performance of blockchain systems has spurred much research. New technology for performance improvement is an area in which researchers have made considerable progress, particularly for permissionless/public blockchains. In this section, some of the performance improvements introduced into Ethereum 2.0 are discussed. Many of these improvements were pioneered in other blockchain platforms and, after much discussion and technical consideration, were introduced into Ethereum 2.0 by the Ethereum Foundation. These performance improvements are an example of the direction in which permissionless/public blockchains are heading.

8.3.2.1 Ethereum 2.0

Starting in around 2017, the Ethereum Foundation sponsored design exercises for architectural changes and open-source development for a series of prototypes designed to improve the performance of the Ethereum blockchain [38]. The goal was to increase the performance from the current average rate of 15 tps up to tens of thousands of tps. The architectural changes can be broken down into the following high-level categories:

- Replace the slow and energy-wasting proof of work distributed consensus algorithm with proof of stake,
- Introduce parallelism into transaction processing by splitting the blockchain into many blockchains, called *shards* that can run in parallel,
- Increase the performance of smart contract execution.

Phase 0 of Ethereum 2.0 was launched on December 1, 2020, with the Ethereum Serenity release. Serenity created a Beacon Chain to coordinate the proof of stake consensus among the various subchains. The Ethereum 1.0 chain continues to run alongside the Beacon Chain and continues to use the proof of work algorithm. Phase 1 introduces sharding and connects the shards with the Beacon Chain, increasing the performance of transaction processing. In Phase 2, ETH accounts, transfers and withdrawals, cross-shard transfers, and contract calls will be added, and Ewasm will be introduced for building execution environments to allow deployment of scalable applications. In Phase 2, the Ethereum 1.0 chain will be linked to Ethereum 2.0, and the distributed consensus algorithm will switch fully to proof of stake.

8.3.2.2 Proof of Stake Consensus

Ethereum 2.0 replaces the slow and energy-wasting proof of work (PoW) algorithm with proof of stake (PoS) [39]. PoS is expected to provide the following benefits:

- Reduction in the amount of energy needed to validate blocks,
- Remove the need for high-performance ASIC-based mining hardware for validation, allowing more people to participate as validators,

- Reduce the Ethereum network's vulnerability to centralization,
- Support sharding since each validator works on a single shard only.

In PoS distributed consensus, a validator must pledge or *stake* ETH in order to act as a validator (32 ETH or roughly $20,000 as of the official Ethereum launch on December 1, 2020). Validators are randomly chosen from the pool to create blocks, and validators that are not chosen, they simply need to check and confirm blocks created by the chosen validator for the current cycle. Validators are rewarded for creating new blocks and for attesting to blocks created by others. If a validator creates or attests to a malicious block, they lose their stake and are removed from the validator pool.

Along with proof of stake, the Beacon Chain has been introduced. The Beacon Chain manages the validators, registering their stake deposits, issuing their rewards, imposing penalties on any validators that misbehave, and coordinating transactions among the shard chains. The Beacon Chain algorithmically chooses a validator on the shard chain to propose new blocks, and the validator is responsible for incorporating transactions into the block. Validators that are not chosen must attest to the validity of blocks, and the attestation is recorded into the Beacon Chain, not in the transaction block itself. At least 128 validators must attest for a block to be incorporated into the chain within a fixed time period. The 128 node validator pool serves for a period of 32 time slots and at the end of that time, the group is replace by a new group. This design reduces the probability that one or several validators will conduct a longer term attack on a shard before being detected. When a block has enough attestations, it is crosslinked from the shard chain into the Beacon Chain and the validator that created the block gets their reward. In addition, the Beacon Chain periodically requires validators to agree to the state of the chain at certain checkpoints, and if two-thirds of the validators agree, the block is finalized.

8.3.2.3 Sharding

Hyperledger Fabric, a permissioned/private blockchain, pioneered the use of parallel blockchains called channels on which collections of organizations or individuals can transact without interfering with each other. Channels were primarily introduced into Fabric as a privacy mechanism since validation, and distributed consensus does not occur between channels. Public distributed ledgers such as Holochain [40] have also introduced sharding, in which each application deployed into the Holochain network maintains a distributed ledger for its own state, rather than having one common ledger for all applications.

Ethereum 2.0 introduces 64 separate shard chains to improve the performance of transaction processing, with the Beacon Chain transferring state information between the shards, thereby allowing all the shards to stay in sync. In theory, this should allow smart contract execution and transaction processing to proceed in parallel, though sharding doesn't eliminate the underlying order–execute–commit architecture on the shards. The nodes performing distributed consensus, formerly known as miners but now called validators since the proof of work algorithm has been abandoned, only work on one shard. Sharding can be combined with rollups to increase performance even further. Rollups allow dapps to bundle transactions into a single transaction off-chain, generate a cryptographic proof, and submit it to the validators for incorporation into the chain, reducing the amount of work the validators must do. Together, these two changes are expected to increase transaction performance dramatically.

8.3.2.4   Faster Smart Contract Execution

In Section 8.1.5, the introduction of Ewasm into Ethereum 2.0 was discussed. While part of the reason for replacing the EVM with Ewasm was the widespread support Wasm has for compilers and other tools and the incorporation of the Wasm language virtual machine into all the major browsers, an additional and equally important reason was to improve the performance of smart contract execution. The EVM was originally designed based on a collection of theoretical principles and not developed with performance in mind, and it hasn't evolved much from the original specification. Because the EVM has to process many different operations, it tends to become an operational bottleneck, and since the bytecode specification hasn't changed, it isn't optimized for different hardware types.

The Wasm compilers can generate much faster code, and they removed Ethereum's dependence on precompiled contracts. Precompiled contracts are like system libraries, but they are bytecodes and not machine code, and are hand-optimized for cryptographic calculations. Without the precompiled contracts, smart contract execution incorporating such calculations would be too expensive to execute. In addition, the widespread support for popular programming languages opens up dapp development to a broader community of developers.

8.4   Summary

In this chapter, we reviewed the basics of distributed ledger technology. Distributed ledgers are built on three separate technological legs: cryptographic hashes and public key signatures, distributed consensus algorithms, and distributed databases. We discussed how these technologies are forged together to form a distributed ledger system and the choices made by the designers of some well-known blockchain systems. When speaking about identity and access management, developers and promoters of blockchain systems tend to speak of permissionless/permissioned and public/private, and we clarified these terms in a way that is more consistent with their general use in identity and access management systems. We then talked about smart contracts and dapps and how they change blockchains from a distributed ledger into a platform for developing and deploying decentralized applications. The transition provides blockchain platforms with a substantially more useful and

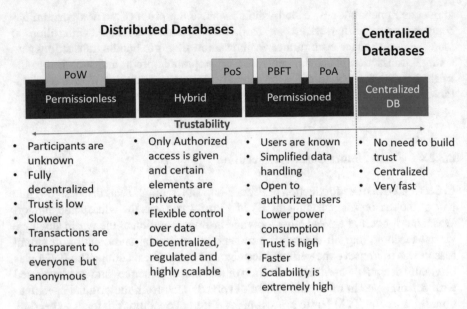

Fig. 8.5 Decentralized vs centralized databases

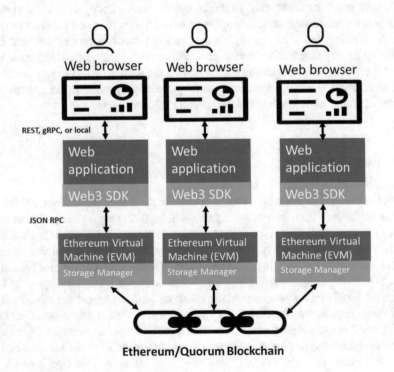

Fig. 8.6 The architecture of a solidity dapp

interesting base of use cases than acting as a distributed database alone. Decentralized identifiers and verifiable credentials are an emerging application of blockchain technology with much promise for simplifying identity management and protecting privacy when verifying authorization for access control, and we discussed the W3C standards and pending standards in this area. Finally, we finished the chapter with a look at Ethereum 2.0, the next generation of Ethereum, which promises to solve the problem of blockchain performance.

Since Bitcoin was introduced in 2009, an enormous amount of energy and research has been conducted in distributed ledger technology. In part, the research was driven by the Bitcoin hype, but while many people equate blockchains and distributed ledger technology with cryptocurrency, the technology has much broader applicability than that. With the advent of Ethereum and its permissioned offshoot Quorum, and the formation of the Hyperledger project at the Linux Foundation to curate new open-source enterprise blockchain projects, a technology base exists for utilizing blockchain in a wide variety of different industries. Wherever there is a multi-party business ecosystem in which the parties depend on a common, distributed database, but have conflicting incentives or don't fully trust each other, a blockchain can help remove the need for trusted third parties that act as pure intermediaries (e.g., they don't provide any value to the ecosystem other than verifying records transferred between parties), which tend to introduce additional cost into the system. Regulated industries also can benefit from distributed ledger technology since the blockchain ledger is immutable after written and constructed in a way to reveal any attempt to tamper with it. All of these properties increasingly hold for the business ecosystem around the generation, transmission, and consumption of electricity. In this chapter, the application of blockchain technology to the electricity industry ecosystem is discussed in more detail.

References

1. W3C, "Decentralized Identifiers (DIDs) v1.0: Core architecture, data model, and representations". [Online]: https://www.w3.org/TR/did-core/ (Accessed 2020-09-01)
2. W3C, "Verifiable Credentials Data Model 1.0". [Online]: https://www.w3.org/TR/vc-data-model/ (Accessed 2020-09-08).
3. Yaga, D., et. al., "Blockchain Technology Overview", NISTIR 8202, National Institutes of Standard In Technology, 2018.
4. Wikipedia, "Bitcoin". [Online]: https://en.wikipedia.org/wiki/Bitcoin (Accessed 2020-09-04).
5. Wood, G., "Ethereum: A Secure Decentralized Generalised Transaction Ledger". [Online]: https://gavwood.com/paper.pdf (Accessed 2020-09-04).
6. Consensys.net, "Consensys Quorum". [Online]: https://consensys.net/quorum/ (Accessed: 2020-09-04).
7. Baliga, A., et. al., "Performance Evaluation of the Quorum Blockchain Platform". [Online]: https://arxiv.org/pdf/1809.03421.pdf (Accessed 2020-09-04).
8. Yumna, G., "Enterprise Blockchains: Quorum — Privacy and Permissioning (part 2)". [Online]: https://medium.com/block360-labs/quorum-privacy-and-permissioning-f7540212211f (Accessed 2020-09-01).

 9. Androulaki, E., et. al., "Hyperledger Fabric: A Distributed Operating System for Permissioned Blockchains". [Online]: https://arxiv.org/pdf/1801.10228.pdf (Accessed 2020-09-04).
10. Ventura, T., "The World's Fastest Blockchain Exceeds 1 Million Transactions Per Second". [Online]: https://medium.com/predict/the-worlds-fastest-blockchain-exceeds-1-million-transactions-per-second-8931df09320d (Accessed 2020-08-31).
11. Rosic, A., "What Is Hashing? [Step-by-Step Guide-Under Hood of Blockchain]". [Online]: https://blockgeeks.com/guides/what-is-hashing/ (Accessed 2020-08-31).
12. Tore, T., "Technical Guide to Generating an Ethereum Addresses". [Online]: https://hackernoon.com/how-to-generate-ethereum-addresses-technical-address-generation-explanation-25r3zqo (Accessed 2020-09-01).
13. Storublevtcev, N. "Cryptography in Blockchain", in: Misra S. et al. (eds) *Computational Science and Its Applications – ICCSA 2019*. ICCSA 2019. Lecture Notes in Computer Science, vol 11620, 2019.
14. Hyperledger, "Channels". [Online]: https://hyperledger-fabric.readthedocs.io/en/release-2.2/channels.html (Accessed 2020-09-01).
15. Ethereum Wiki, "Ethash". [Online]: https://eth.wiki/en/concepts/ethash/ethash (Accessed 2020-09-07).
16. University of Cambridge, "Cambridge Bitcoin Electricity Consumption Index". [Online]: https://www.cbeci.org/ (Accessed 2020-09-07).
17. Lihu, A. et. al., "A Proof of Useful Work for Artificial Intelligence on the Blockchain". [Online]: https://arxiv.org/pdf/2001.09244.pdf (Accessed 2020-09-07).
18. Costan, V., and Devidas, S., "Intel SGX Explained", [Online]: https://eprint.iacr.org/2016/086.pdf (Accessed 2020-09-08).
19. Ethereum Foundation, "Ethereum 2.0 Specifications". [Online]: https://github.com/ethereum/eth2.0-specs (Accessed 2021-02-24).
20. Wikipedia, "Byzantine fault". [Online]: https://en.wikipedia.org/wiki/Byzantine_fault (Accessed 2020-09-07).
21. Coindesk, "Writing Bitcoin Smart Contracts Is About to Get Easier With New Coding Language". [Online]: https://www.coindesk.com/bitcoin-smart-contracts-minsc-easier-new-coding-language (Accessed 2020-09-03).
22. Readthedocs.io, "Solidity – Solidity 0.7.1 documentation". [Online]: https://solidity.readthedocs.io/en/v0.7.1/ (Accessed 2020-09-03).
23. OAI, "OpenAPI 3.0 Specification". [Online]: https://github.com/OAI/OpenAPI-Specification/blob/master/versions/3.0.0.md (Accessed 2020-09-03).
24. Ethereum Foundation, "Ethereum flavored WebAssembly (ewasm)". [Online]: https://github.com/ewasm/design (Accessed 2021-02-24).
25. WebAssembly, "WebAssembly Specifications". [Online]: https://webassembly.org/specs/ (Accessed 2021-02-24).
26. JSON RPC Working Group, "JSON-RPC 2.0 Specification". [Online]: https://www.jsonrpc.org/specification (Accessed 2021-08-22).
27. Readthedocs.io, "Web3.py 5.12.1 documentation". [Online]: https://web3py.readthedocs.io/en/stable/ (Accessed 2020-09-03).
28. Readthedocs.io, "web3.js – Ethererum Javascript API". [Online]: https://web3js.readthedocs.io/en/v1.2.11/ (Accessed 2020-09-03).
29. Vogelsteller, F., and Buterin, V., "EIP-20: ERC-20 Token Standard". [Online]: https://eips.ethereum.org/EIPS/eip-20 (Accessed 2020-09-08).
30. Entrinken, W., et. al., "EIP-721: ERC-721 Non-Fungible Token Standard". [Online]: https://eips.ethereum.org/EIPS/eip-721 (Accessed 2020-09-08).
31. Radomski, W., et. al., "EIP-1155: ERC-1155 Multi Token Standard". [Online]: https://eips.ethereum.org/EIPS/eip-1155 (Accessed 2020-09-08).
32. Json-ld.org, "JSON for Linking Data". [Online]: https://json-ld.org (Accessed 2020-09-08).
33. W3C, "DID Specification Registries". [Online]: https://www.w3.org/TR/did-spec-registries/ (Accessed 2020-09-08).

34. Decentralized Identity Foundation, "DIF Identity Hubs". [Online]: https://github.com/decentralized-identity/identity-hub/blob/master/explainer.md (Accessed 2020-09-08).
35. Wikipedia, "Bitcoin Cash". [Online]: https://en.wikipedia.org/wiki/Bitcoin_Cash (Accessed 2021-02-25).
36. Visa, "Visa Fact Sheet". [Online]: https://usa.visa.com/dam/VCOM/download/corporate/media/visanet-technology/aboutvisafactsheet.pdf (Accessed 2021-02-26).
37. Gorenflo, C., Lee, S., Golab, L., and Keshav, S., "FastFabric: Scaling Hyperledger Fabric to 20,000 Transactions per Second". [Online]: https://medium.com/interdax/ethereum-2-0-explainer-e996ac7dc006 (Accessed 2021-02-26.
38. Consensys.net, "Ethereum 2.0 FAQ". [Online]: https://consensys.net/knowledge-base/ethereum-2/faq/ (Accessed 2021-02-26).
39. Wakerow, P., "Proof-of-stake (PoS)". [Online]: https://ethereum.org/en/developers/docs/consensus-mechanisms/pos/ (Accessed 2021-02-26).
40. Holochain.org, "Holochain". [Online]: https://holochain.org/ (Accessed 2021-02-26)

Chapter 9
Energy Systems Meet with Blockchain Technology

Renewable energy power generation from both utility-scale plants and distributed energy resources is replacing traditional centralized generators powered by fossil fuels in national electrical grids around the world. This trend is driven by global and national goals to decarbonize national grids as a response to global climate change. In the energy transition toward decarbonization and decentralization, the power industry has been deregulated, and energy markets have been liberalized in many countries. The development and deployment of clean energy resources have been promoted by both clean energy regulatory policies and energy market mechanisms. In addition, the cost of renewable energy system components has dropped dramatically, and energy conversion efficiencies have improved as economies of scale in manufacturing and the basic technology have advanced.

When renewable energy devices such as solar panels, wind machines, and batteries for storage are deployed closer to the load sites, power losses over long-distance transmission lines can be reduced as compared to fossil fuel-powered generation plants located far from the load. In many cases, the cost of renewable energy devices is affordable enough that traditional energy consumers are transforming to prosumers, entities that produce energy through renewable devices and consume it in electrical loads. For example, residential houses, traditionally viewed as energy consumers, can now also become energy producers with the installation of roof-top solar PV panels. The prosumer deployment also applies to any type of distributed generation and energy storage deployed in a smart building or electric vehicle, as well as to load flexibility, where a consumer can reduce their load in response to an automated request from the distribution system operator.

Prosumers pose challenges for the current power market structure. The structure of the power market on a national or regional level does not consider such small-scale energy resources or the local energy balance. In addition, wholesale markets are not able to deal with the almost real-time intermittency and stochastic nature of renewable energy resources at high precision [1]. Future energy markets need to integrate prosumers, small-scale load flexibility, and local energy generation. New local energy market mechanisms need to be deployed. New market structures for power markets, including local market rules and pricing mechanisms, are among the emerging research topics in academia. Recent advances in information and communication technologies (ICT), data analytics, control, and optimization algorithms play a key role in supporting this transformation. Two important problems that need to be solved in order to realize the decentralized energy markets of the future are tracking energy supply and demand in real time or near real time and coupling that tracking with charging and billing transactions.

Blockchain distributed ledger technology (DLT) is one of the most promising technologies for solving these problems. DLT promises to play a key role in the decentralization of the energy industry and in future decentralized power markets. Blockchain DLT features decentralization, democratization, and transparency, important characteristics for ensuring the viability of the future decentralized power systems and markets (see Fig. 9.1). DLT facilitates energy transactions, smart contracts, and instant transfer of charging and billing records in a reliable way. This chapter provides a systematic overview of DLT use cases in the energy industry.

9.1 Energy Blockchain Use Cases

Blockchain DLT can be applied to numerous types of grid and power market applications. However, most use cases are still being developed and refined. Based on a review of the literature, the most prominent use cases of blockchain in the energy section are identified as follows: [2].

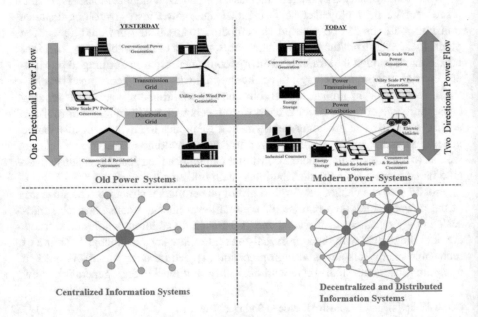

Fig. 9.1 Decentralization of power systems and distributed ledger technology

- Labeling, energy provenance and certification,
- Decentralized and peer to peer (P2P) energy trading,
- EV charging and eMobility,
- Wholesale trading and settlements,
- Flexibility management,
- Energy financing and ICO crowdfunding,
- Sustainability (REC trading, etc.),
- Others.

Reference [3] presents a segmentation of blockchain implementations in the energy sector focusing on the type of use case and applications, such as P2P energy trading and transactions, grid operations, and EVs. However, the use case segmentation contains very little operational and technical detail. Little information is given about the specifics of the blockchain technology for each segmented use case; hence, it is not straightforward to distinguish the specific advantages of the discussed implementations over more traditional technological approaches. DENA (Deutsche Energie-Agentur) demonstrated another energy DLT segmentation method in which clustering is done according to main groups such as trading, communication, marketing, financing, data, and asset management [4]. The segmentation in the DENA study has also limited details and is therefore of limited use for developing a systematic case for deploying blockchain in energy markets. However, references [5, 6] propose an alternative and systematic methodology to demonstrate the value of blockchain DLT in various energy DLT use cases in power systems and energy markets.

In Fig. 9.2, the value chain for the entire energy market is summarized, from generation, transmission, distribution, to retail and prosumer/consumer segments. Renewable energy systems (RESes), electric vehicles (EVs), and energy storage systems (ESSes) can be integrated into any segment depending on the scale and functionality. Small-scale distributed solar and wind resources can be integrated into the distribution system, and large-scale solar/wind farms (onshore or offshore) can be connected to the transmission system. Utility-scale energy storage can be connected to transmission networks, and community or smaller-scale energy storage can be connected to distribution power grids. Behind-the-meter (BTM) solar and energy storage systems are connected through prosumers into the low-voltage distribution grid.

The aim of the segmentation presented in Fig. 9.2 is to provide a better understanding of the DLT-based use cases by mapping out the existing applications versus the power market and systems segment where the use case occurs. The segmentation methodology is based on four characteristics:

1. Data recording,
2. Financial transactions,
3. Energy transactions,
4. Smart contracts,

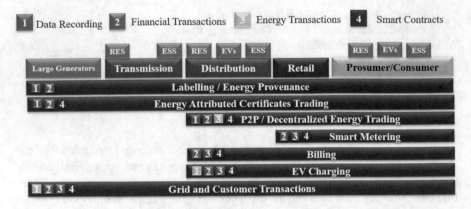

Fig. 9.2 Energy DLT uses case segmentation

Data recording is a fundamental characteristic of all DLTs since they are responsible for storing digital records in a form that is immutable and transparently available to all participants with permission to read the ledger. In energy DLT use cases, all types of operational data can be stored in a DLT. In principle, all DLT-based energy use cases utilize data recording. The recorded on-chain data provides a certain level of cybersecurity for the recorded features and provides the basis for specific use cases such as labeling/energy provenance, EV-related operations, and proof of origin for renewable energy traceability.

The financial transactions characteristic converts the commodities, services, tangible, and nontangible assets into digital entities via a tokenization process. The tokenized value of the transacted entity is basically a digital tracking record that represents the value of that entity. The value record can be kept in the form of a blockchain-based cryptocurrency or simply as the digital equivalent of the fiat currency that is valid for the country in which the power system for which the DLT is deployed is located. Energy transactions represent electrical energy in some form, sometimes as an energy token, in DLT records so that its transfer and utilization by loads can be tracked. Smart contracts are the digital counterparts of the paper-based contracts that are used to govern the exchange of electricity between participants in the energy market and between participants and end consumers. Depending on the use case, smart contracts can be the ultimate enablers of energy DLT-based businesses in which certain rules are set between the transaction parties.

The primary application of blockchain DLT currently is in transactive P2P retailer electricity and energy trading platforms for residential, enterprise-campus microgrids and municipal facilities. Deployments are still in the early proof-of-concept (POC) stage. A number of emerging start-up companies are developing technical solutions, services, and products for these applications. The remaining blockchain-based grid applications are mainly focused on the mission-critical grid operational side and will be discussed in the following sections.

On the demand/customer side, typical retailer or prosumer applications include peer-to-peer energy trading, transactive control, electric vehicle (EV) charging management, demand response, energy efficiency, smart home energy management, and smart metering and billing. On the grid operations side, system operators are investigating blockchain DLT for end-to-end regulatory compliance, grid asset management, and renewable energy certification (REC) at the point of provenance in utility-scale solar and wind farms. Both types of use cases are discussed in the remaining sections.

9.1.1 Labeling, Energy Provenance, and Energy Attribute Certificates

Decarbonization has been a global goal for energy supplies for the last two decades. Governments and companies set ambitious sustainability targets, but accomplishing such ambitious targets requires more investment in renewable energy technologies and the development of an improved method of labeling and tracking the origin of generated electricity. Tracking the origin of electricity, or energy provenance is accomplished by various tracking and certification systems such as Renewable Energy Certificates (RECs) in the US and Canada, Guarantees of Origin (GOs) in Europa and Russia, International Renewable Energy Certificates (I-RECs), and national systems in the rest of the world. These certificates are designed to provide evidence of energy provenance per MWh. A general term for the certification of energy provenance is Energy Attribute Certificates (EACs). EACs essentially unbundle the energy from the renewable attributes, allowing the renewable attributes to be traded or sold, while GOs allow end customers to select their electrical energy from a particular supplier that generates from renewable sources.

EACs allow an energy producer with excess renewable energy to sell the renewable attributes separately from the actual energy so that a distribution system operator, which is short of the mandated renewable content in the energy they supply to customers, can purchase the EACs to meet the mandate. Once the EACs have been purchased, they are retired and are no longer for sale. EACs are not used to track the flow of energy but rather as a kind of accounting instrument. The majority of existing EACs are based on annual green accounting systems that track the average production of green energy and its consumption. The EACs can be dynamically traded on markets as nontangible assets in several jurisdictions, but can also be traded bilaterally. An EAC has a certain lifecycle from its origination and registration, a certification that makes it a tradable asset to the storage value of the intangible asset (certificate), and the exchange through to its retirement or usage.

In principle, an EAC can be bought from any location and time; therefore, it is also essential to record the origin of the generated green electrical energy (each kWh energy block), the timestamp (time of energy generation), and the location of the power generating asset. EACs allow large companies or organizations to decarbonize their power supply without necessarily owning their own renewable energy generating equipment or buying renewably generated electricity directly. Under normal conditions, the final owner of the EAC is supposed to cancel or retire the traded certificate once it is consumed to prevent double-counting, which can even be considered a fraudulent activity if engaged in. Since transactions for EACs using existing platforms may involve multiple agents, the ability of the system to prevent fraud may be limited. The double-counting problem is a known challenge for current EAC platforms. Besides, some of the processes for validation and audit of the certificate status require manual updating by central authorities in many national systems (Fig.9.3) .illustrates the process used by existing platforms in comparison with a DLT-based platform.

Despite the fact that most EAC systems are now based on Web technology, they lack secure, cost-effective, reliable, efficient, and fast solutions. Many steps are manual and involve transferring documents from one Web site to another, require a person to approve the document, and then require the person to deliver the document by email or file download to the purchaser. Hence, several developers and consortiums such as Energy Web Foundation (EWF) have started to offer blockchain DLT-based platforms for EAC trading as well as other certifications in other emerging green attribution markets such as carbon credits/taxes [7]. A DLT-based EAC system can accommodate a central digital registry and certification authority that receives the origin of provenance, the digital ID of the green energy-producing asset, location of the asset, and timestamp of the energy generation in the initial round before the system approves a new EAC. Once the EAC is generated, the status of the approved certificate is recorded along with the EAC traders' activities. A trader may prefer to re-trade (swap) the EAC or consume it. If it is consumed, the latest status of the EAC is then labeled as "retired" to eliminate the double-counting problem and any associated fraudulent activities. DLT-enabled EAC trading platforms may improve the overall system performance and reduce the costs associated with third parties, reduce longer transaction times, and prevent possible fraudulent activities.

DLT-based market platforms are already deployed for EAC origination, trading, and consumption. Large companies like Google and Microsoft have already declared their sustainability targets, which aim to use 100% green energy for their operations [8, 9]. Microsoft initiated a showcase project with Vattenfall to develop EAC

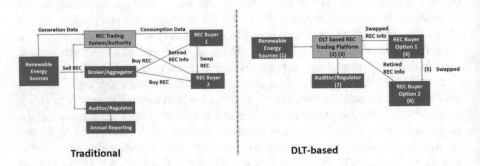

Fig. 9.3 REC trading with and without DLT

solutions in Sweden. The Australian startup company Power Ledger is offering a carbon trading solution using blockchain technology [10]. They partnered with European green energy provider companies such as ekWateur in France and Alperia in Italy to offer a blockchain-based EAC platform to their customers [11, 12]. The Thai state-owned energy company PTT announced a blockchain-based renewable energy marketplace using I-RECs with the Energy Web Foundation [13]. In addition, SP Group created a DLT-based I-RECs trading platform in Singapore [14].

9.2 Decentralized and P2P Energy Trading

With the increasing integration of distributed energy resources (DERs) and smart meters into the power grid, conventional passive consumers are becoming more active in managing their own consumption and can, in principle, participate in grid operation and management. Increasing coordination between DERs, energy storage systems, and the grid as a whole has the potential to increase the economic viability of DERs and batteries and offer additional power systems services. Transactions may occur in a decentralized and or even a P2P manner depending on the application. The surplus energy generated or stored by prosumer devices can be traded with their immediate neighbors in a P2P manner in a microgrid. Decentralized transactions don't necessarily need to use P2P methodology but can also use routed methods, for example, storing the surplus in a bulky energy storage unit to be discharged to the demand side depending on the need. These new technologies are driving a restructuring of the current energy markets toward a bottom-up structure, enabling end-prosumers to actively participate in energy markets. Such transitions can be enabled by DLT, which avoids the introduction of a third party for managing the energy transactions and preserves the privacy of consumers. DLT-based platforms can also offer smart contract features to minimize the on-paper contracts between the transacting parties. The rest of this section summarizes several pilot projects that utilize blockchain and tokens to develop a P2P trading platform for prosumers and consumers in microgrids and communities.

LO3 Energy LO3 Energy was one of the earliest companies to develop a blockchain-based platform for grid applications and the energy industry. Their P2P energy trading platform was deployed on a microgrid in Brooklyn, New York, in 2015. Their projects since have expanded to other parts of the US in addition to Australia, Europe, and Japan. Their DLT-based transactive energy platform, "Pando," allows energy to be traded among utilities, retailers, prosumers, and consumers in a secure, customized, and scalable way. Until recently, LO3 was using a private ledger named Exergy with XRG tokens. Exergy supports a global private and permissioned blockchain. A data platform now is also provided by Exergy, where uses can access real-time and historical data from behind-the-meter resources to learn market behaviors and bidding strategies. On September 20, 2020, LO3

Energy announced that Pando will migrate to the Energy Web Chain blockchain platform [15].

Power Ledger Power Ledger is an Australian developer of blockchain-based trading platforms for P2P energy trading. Until recently, Power Ledger used the Ethereum blockchain platform in addition to a private blockchain called EcoChain, but recently announced that they will be moving to the Solana platform, which utilizes the less energy intensive PoS consensus protocol. The company has signed an agreement with the Thai energy company BCPG Group to create a digital energy business enabling P2P energy and environmental commodity trading across Thailand. Power Ledger supports dual-currency systems, which allows the use of the local fiat currency and a private token called Sparkz-based on EcoChain. Sparkz can enable the use of fiat currencies in national markets using a closed-loop exchange. Power Ledger also supports an ERC-20 compatible token, POWR, which is publicly tradable on cryptocurrency exchanges. These two tokens are deployed in a dual manner for market flexibility and operate through two blockchain-based market products. Sparkz is recommended for local energy market applications and POWR for wholesale energy market transactions. Public transactions of POWR tokens are made on the public Ethereum blockchain, while private transactions are made on Power Ledger's own consortium blockchain EcoChain, which uses the proof of stake (PoS) consensus protocol [16].

Greeneum Greeneum offers a blockchain and personalized AI platform that is designed to accommodate various energy trading operations. Energy data is confirmed through a validation process. The validated data is recorded on the Ethereum blockchain as an immutable record. The system runs periodic calculations of production and consumption on the grid and enables consumers to directly pay producers using GREEN Tokens. Producers of green energy are rewarded with additional tokens, called GREENEUM bonds, and receive Greeneum carbon credits in exchange for their user's consumption of green energy [17].

Verv Verv Energy is a British company launched in 2018 that offers a blockchain-enabled P2P energy trading platform based on high-speed data acquisition and AI technology. The platform can be used to understand the consumption and production of local energy. In April 2018, they claimed to have conducted the UK's first P2P trade of energy on a blockchain. Verv still faces some regulatory challenges to moving their system from a technology prototype into full scale production. They use VLUX tokens [18].

Energy21 & Stedin Energy21 & Stedin launched a pilot in 2019 in cooperation with ABB, i.LECO and Stedin (a DSO) in the Hoog Dalem district of Gorinchem, the Netherlands. Fifteen residents are taking part as prosumers/consumers. Participants are able to see their own consumption/production and the amount traded locally through a blockchain-enabled application. Their platform uses a federated Byzantine agreement consensus protocol and the QUASAR blockchain.

QUASAR is a permissioned consortium blockchain technology that was developed by QUANTOZ in 2015 in the Netherlands [19]. The purpose is self-consumption or the mutual exchange of electricity within the community accompanied by immediate price settlement. When self-production falls short, energy can be bought from the grid using flexibility or ancillary grid services. QUASAR also includes data exchanges with sensors in the local grid for interactions with smart home systems [20].

9.3 EV Charging

EV charging has many attributes of a promising use case for DLT implementation. In this section, DLT-based solutions for EV charging are briefly reviewed. In some cases, companies may provide other services related to mobility, such as toll road and parking payment, implemented with blockchain technology, but we focus here specifically on EV charging solutions.

Mobility Open Blockchain Initiative (MOBI) MOBI [21] is an industry consortium including a broad spectrum of automobile manufacturers, insurers, IT companies involved in cloud and blockchain, etc., that is working toward standards for connected vehicles. While full access to their standards documents is only available to members, they have released a summary of their published standards in the areas of vehicle identity management (VID), electric vehicle to grid integration (EVGI), and the connected vehicle marketplace (CMDM). The CMDM standard specifies a blockchain as the basis for communicating data between a vehicle and some other entity, like an EV charger. The standard describes a reference architecture, data schema, certificates, and other important data types for vehicles to exchange information with infrastructure through a blockchain. The VID standard builds on the current Vehicle Identification Number (VIN) to provide a digital identity certificate for the vehicle. Each vehicle is provisioned with a "birth certificate" when it is manufactured that can be used to identify the vehicle to any infrastructure device. The EVGI standard covers the high-level system design, reference architecture, multiparty processes, and data schema that are necessary to implement a vehicle to grid (V2G) integration. It also provides guidance for implementation and discusses three use cases. It does not, however, specify any specific technologies for implementing the standard, for example, what blockchain technology to use or even what protocols to use at layer 2, layer 3, and above. While the EVGI standard is focused on EV charging, the other standards developed by MOBI, while necessary for V2G integration, have broader applicability in the transportation sector.

Energy Web Foundation The Energy Web Foundation (EWF) has been working with charging network services and automobile manufacturers to enable charge network roaming on the Energy Web Chain, an open-source, enterprise blockchain platform designed for the energy market. The Share&Charge Foundation has implemented its Open Charging Network (OCN) on top of the Energy Web Chain [22]. Share&Charge is used by eMotorWerks and other charging network operators to allow EV drivers to utilize charging stations run by charging network providers other than the provider they signed up with (their home provider). OCN uses the Open Charge Point Interface (OCPI v2.2) protocol discussed in Section 5.5.4.3 between the EV and charging station. OCN implements the OCN Registry [23] through a smart contract on the Energy Web Chain. When an EV driver initiates a charge session at a charging station enrolled in OCN, the charging station checks the blockchain registry to route the charging event to the driver's home network for settlement.

Research Solutions There have been many research papers published over the last 10 years with architectures and prototypes of EV charging using DLT technology. Many focus specifically on how to increase the use of electricity with renewable content in EV charging networks. For example, [24] proposes to allow local renewable energy providers such as building owners with solar PV to supply public charging stations with energy. When an EV needs to charge, the driver via a smartphone app or an autonomous agent on the EV triggers an auction through a smart contract running on a private Ethereum network. Both the EVs and the renewable energy providers have smart meters to report on the amount of energy produced and consumed. The system relies on the billing infrastructure of the distribution system operator and requires the distribution system operator to additionally manage the smart meters for the local energy providers.

9.4   Grid and Customer Transactions

Integrating Energy IoT with DERs and smart meters allows customer-owned, behind-the-meter resources to participate in energy programs, including grid stabilization programs such as flexible load management, through bidirectional communication between the DSO and the DERs and real-time recording of net energy use. Real-time wholesale and retail energy markets, flexibility management, and associated services at both the DSO and TSO level are other high potential use cases for DLT-based platforms. The rest of the section describes some applications of DLT to these use cases.

Aizu Laboratory (AizuLab) is a Japanese-based start-up company that originated from Aizu University in 2007. The company focuses on developing hardware and software utilizing cutting-edge technologies in several fields, including energy. It is has partnered with Eneres, a Tokyo-based company providing electricity-related services, for testing the capabilities of distributed registry technology. The test involves smart microgrid networks consisting of renewable energy resources with about 1000 households in Fukushima and investigates the possibility of building a shared energy economy. The test indicated emphasized the importance of a "sharing economy" approach in communications with customers and investigated block-chain technology for smart grid application, in which households equipped with solar and other power generation systems supply surplus electricity to other house-holds while receiving a reward [25].

Drift is a US-based start-up energy retailer supporting peer-to-peer trading in retail energy markets for small-sized customers and generators [26]. The energy trading platform developed by the company provides an environment to be able to monitor transactions for participants. Drift promises their customer to purchase the energy from 100% renewable resources and constantly interacts with generators to build a daily supply curve matching generation and demand, then purchases the energy for their customers once the desired supply curve is generated. The strategy the com-pany uses is similar to high-frequency trading to reduce or eliminate any price spikes. Drift also aims to conduct fast transactions by reducing middlemen, fees, and bureaucracy during energy trading.

Quantoz is a Dutch start-up company founded in 2014. It specializes in building and implementing blockchain-based solutions, such as NEXUS and QUASAR, to harness the power of blockchain technology and cryptography. NEXUS provides solutions for automated token, crypto, and fiat transaction processing, while QUASAR enables instant and compliant peer-to-peer (micro) transactions between enterprises, people, and IoT devices. These solutions also fully address the gap between the integration real-world applications with blockchain technology. The company developed a blockchain concept called the Layered Energy System (LES), which enables a decentralized and layered energy-sharing economy for local energy communities. Quantoz has also started a pilot project with its partner Energy21, an expert in energy and data management solutions, for the Dutch grid operator Stedin [27]. The grid operator has been able to further explore the concept successfully and implemented a pilot project for a local energy market in a Dutch community. The partners plan to further develop the blockchain platform and use it for other projects to expand the concept from local energy markets to local grid capacity markets.

Sunchain is a French start-up using blockchain to develop smart energy solutions for applications such as collective self consumption, certification, and green mobil-ity. It focuses on applying blockchain-based solutions to new use cases of solar energy, such as managing local power exchanges between prosumers and consum-ers [28]. The company developed a virtual blockchain-based network for solar energy prosumers, which allows them to efficiently deal with energy transactions at low cost and manages energy exchanges between parties [29]. Their secured virtual network can also track and certify the energy flow in the power grid, identifying produced and carried energy, for example, and identify the value of surplus energy in the neighborhood, in addition to providing services for EVs, for example, charg-ing an electric vehicle remotely. The network consists of several nodes embedded on IoT devices, which record data from smart meters into the blockchain.

Alectra Utilities is one of Alectra's subsidiaries (the second largest municipally owned integrated energy solution company in North America) and is responsible for distributing electricity to residents and businesses in Canada. The company developed a blockchain-based transactive energy platform to facilitate energy transaction management between wholesale, distribution, and end users. The platform allows Alectra to request power from customers. They receive compensation through a virtual currency that has been created to support participation. The currency can be exchanged for goods and services at participating merchants or for other forms of value in a marketplace [30], like frequent flyer points offered by credit card providers. It has also collaborated with Interac, a payment processing company, to develop a marketplace using DLT that provides a single, permanent ledger enabling energy transactions to be settled and digital assets to be exchanged among all stakeholders. The DLT system is based on IBM's Hyperledger framework [31].

9.5   Metering and Billing

Electricity usage monitoring is an important function for providing an instant view of active energy usage in the metering and billing applications. However, the privacy of personal data and the user's trust in the smart grid system needs to be considered when implementing metering and billing applications. Blockchain enabling distributed ledger technology (DLT) can provide safer and more transparent solutions due to its decentralized structure. DLT solutions primarily deal with the following use cases: development and usage of smart meters for peer-to-peer (P2P) energy trading, electricity invoice settlement via cryptocurrencies such as Bitcoin, and automated CO_2 impact calculations based on energy mix profiles. The rest of the section describes some implementations of services addressing these use cases.

Bankymoon is a South Africa-based startup working on a blockchain-based solution, especially to enable prepaid utility meters. Prepaid meters have a digital currency wallet that can be reached from all over the world. Consumers can buy electricity, gas, and water easily without going through a bank. Also, blockchain-aware smart meters can be installed in schools to receive donations from the international donor market [32]. Their latest project is the development of smart electricity meters that can work with the Bitcoin network for smart P2P energy trading.

BAS Nederland is a Dutch energy supplier and stands for "Beneficial to All Stakeholders." In 2018, it became the first energy company around the world to accept Bitcoin as payment for bill settlement [33], allowing the company's customers to pay their electricity bill with Bitcoin [34]. The company started a project titled "De Weg Nar Nul," which means "Road to Zero," to help its customers to progress toward complete energy independence, that is, customers will be able to provide energy for themselves in the future.

Elegant is a Belgium-based green energy supplier founded in 2012, which supplies electricity and natural gas to almost 30,000 households. Elegant started to accept Bitcoin payments for energy services, including gas and electricity, following requests from various customers [35]. It has partnered with BitPay to process payments in digital currency. The company positions itself as a local green energy supplier, and incentives are given to renewable energy sources. The company does not use any blockchain technology and only accepts invoice settlements with Bitcoin through its partner, that is, BitPay.

Enercity is one of the 10 largest municipal utility companies in Germany and offers energy to Hannover, Niedersachsen, and some nationwide deliveries in "key account areas." It allows its customers to pay their bills with Bitcoin. Enercity is one of the first energy companies in Germany to introduce the possibility of paying with Bitcoin [36]. The exchange from Bitcoin into Euro is automated, and the billing processes are managed by the service company Pey GmbH. The company does not use any blockchain technology itself.

Engie is a French-based company operating in the energy sector. It collaborates with a company called Ledger [37], which is a secure crypto-wallet company. Ledger developed the first blockchain-based hardware secured data storage at the source of energy production, that is, a smart meter enabled with tamper-proof blockchain technology for data integrity. The device also measures the data at the source of green energy production, such as wind turbines, solar panels, or hydropower, and records it securely into the blockchain [38]. The energy production and use information is registered in tamper-proof certificates on the blockchain and accessible to users via their software called TEO (The Energy Origin).

9.6 How to Build DLT-Based Services in the Energy Domain

The development of DLT-based services for energy use cases requires a multidisciplinary team with know-how in software development, economics, regulatory issues, energy systems, and markets. A DLT-based development effort starts with comprehensive technology scouting and market analysis as the first step. In this step, it is essential to investigate the features and main functionality of possible blockchain DLT options. In addition, a detailed market analysis of the targeted market segment is critical. The investigation should include but not be limited to gap, competitor, trend, and growth analysis sections. Once the first step is accomplished, the results should be used to narrow down and identify the use case details that then need to reflect the business logic. Any regulatory concerns related to the use case should be understood up front since failure to abide by the regulatory regime of the jurisdiction in which the business will be conducted could collapse the entire business operation. For example, a P2P energy trading service cannot be operated where such power transactions are not allowed by the regulatory regime (Fig. 9.4).

The third step is determining whether DLT is the right technology for the use case. As discussed in Chapter 8, there are a particular set of criteria that apply to use cases which are a good match for DLT technology, and if your use case does not satisfy these criteria, then it might not be a good candidate for implementing with a blockchain. If the investigated use case requires distributed databases and digital ledgers to eliminate unnecessary third parties or to reduce the number of intermediaries by building trust and enabling cybersecurity, the use case is a good candidate for applying DLT. The following list expands on the list of general criteria described for good blockchain use cases from Chapter 8 and applies them specifically to the energy domain (Fig. 9.5):

1. Do you need an intermediary for your business or use case or can you eliminate them?
2. Are your assets already digitalized, or can you digitalize them?
3. Are high-frequency transactions required?
4. Do you need to store bulky non-transactional data?
5. Do you require a shared and distributed database?
6. Are value exchange and contractual relationship management an integral part of your business?
7. Is the control functionality required?
8. Can all transaction data be public?
9. Can consensus be determined in the intra firm network?

Preparing a business plan is another very important milestone of the project. A well-structured business plan covers the following:

- Executive summary of your project
- Description of your team (cofounders, leadership team, personnel, and consultants),
- Description of the proposed DLT-based product or service,
- Summary of 1–3 included technology scouting, market analysis, use case identification and eligible DLTs for the use case,
- Sales, marketing, distribution, and strategic partnership planning,
- Management plan for the coworkers and business coordination,
- Analysis of legal and regulatory issues,
- Comprehensive strengths, weaknesses, opportunities, and threats (SWOT) analysis,
- Financial planning and sales projections are supported by the projected market size.

Prototyping and proof of concept (PoC) development steps are the first practical and hands-on parts of the process. Figure 9.6 demonstrates a sample architecture of a generic DLT-based implementation. The back-end of the generic architecture has two sections: a blockchain layer (on-chain domain) where the real blockchain technology, such as Ethereum or Hyperledger Fabric, is deployed. All other domains, excluding the on-chain domain, are defined as off-chain domains of the service. The second part of the backend can host software development kits (SDKs). The front-end is the application layer where the graphical user interface (GUI) is implemented. Business logic is integrated in various sections of the architecture. Business logic reflects the business rules and routines of the target use case and peripheral components. ML/AI-based algorithms, optimization algorithms, controllers, cloud services, file/data storage systems, and other necessary components can also be added to the developed system depending on the use case and application details. Performance measurement is important to ensure that the application meets user requirements and should be undertaken when the PoC code is complete.

The last step is the commercialization and full-volume production, where the product managers, software developers, and other technical team members come up with the alpha (prerelease) version of the service while the sales and marketing team focuses on how to market the developed service to potential customers.

9.7   Performance and Scalability

Several blockchain frameworks have been developed that provide flexible and adaptable platforms including Hyperledger Fabric, Corda, Ethereum, IOTA, Omni, OpenChain, Ripple, MultiChain, and Chain Core. Although there are several blockchain projects in the energy domain that have been implemented on these platforms, technical challenges and concerns regarding blockchain platforms still remain. The challenges involve latency, scalability, and throughput [39].

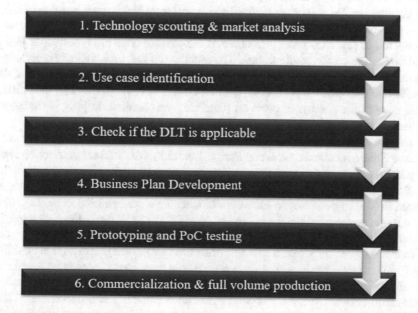

Fig. 9.4 DLT-based implementation development steps

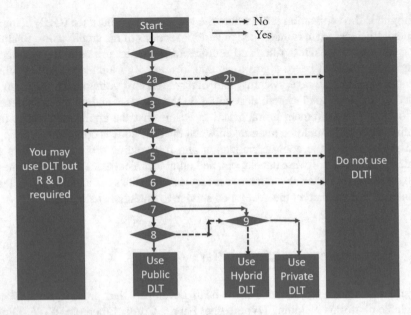

Fig. 9.5 Do you need DLT for your business?

In [40], the performance of the Hyperledger Fabric platform was evaluated and discussed. Two versions of Hyperledger Fabric (v0.6 and v1.0) were compared in [41], and the results demonstrated that the performance of Fabric v0.6 decreased with an increase in the number of nodes; in contrast, the performance of v1.0 was not impacted by the number of nodes. In [42], different block sizes, number of channels, endorsement policies, resource allocation, and state database choices were considered to investigate the difference in the performance of Hyperledger Fabric v1.0 depending on the configuration. The authors of [43], investigated Hyperledger Fabric v1.1 and the impact on the latency and throughput of varying the block size (i.e., size of transactions per block), peer CPU, and SSD vs. RAM disk. In addition, the impact of the number of peers on scalability was studied. In [44], performance metrics were recommended for assessing Hyperledger Fabric.

The performance of the Ethereum blockchain was evaluated in [45] for most popular Ethereum clients, Geth and Parity. The impact of different transaction types on performance was also investigated. The performance evaluation results show that the Parity client can support faster transactions than the Geth client. A test was conducted in [46] to evaluate the Ethereum blockchain's performance in reading/writing data on a relational database such as MySQL. In [47], the authors discussed a prototype blockchain network that was implemented to store the health information of individuals and reported a low response time (<500 ms), and high availability (98%) for the use case analyzed. In summary, the throughput, latency, and scalability of an Ethereum blockchain network depend on the hardware configuration, blockchain network design, and smart contact complexity and operations. These results may differ depending on the test environment. The findings reported in this

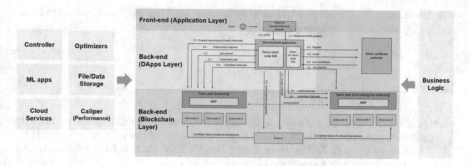

Fig. 9.6 Sample architecture of a DLT-based implementation

papers can serve as a guideline that can help select a suitable hardware configuration, as well as a blockchain network and its parameters that can support a specific blockchain implementation and requirements.

Blockchain services deployed in a cloud are another option for building an energy domain service. In [48], up to 200 TPS are reported using AWS EC2 with VMs having 16 vCPUs, running on a 3.0 GHz Intel Xeon Platinum processor with 32GB RAM. The experimental DLT testbed can accommodate 100,000 participants. Since the performance of the DLT platform depends on the number of participating nodes, total data stored in the system the consensus mechanism and many other parameters are related to scalability.

9.8 Summary

Power systems and markets are evolving and are rapidly becoming more decentralized, decarbonized, and digitized quite rapidly AI/ML and blockchain technology are among the digitalization technologies having the most impact and which can accommodate complex interactions such as bidirectional energy flows and distributed computation. In this chapter, the use of blockchain technology in the field of power systems and markets is discussed. Since blockchain DLT technology has the potential to transform businesses in a disruptive manner, it is essential to explore the real potential of this emerging technology and initiate healthy discussions beyond the initial hype created around blockchain, fueled by its association with the initial use case of cryptocurrency. P2P energy trading, EV charging, and green energy certificate trading appear to be the most well-developed use cases currently. Even though the current generations of DLT and blockchain technologies have great momentum, they still have some issues related to interoperability, scalability, and performance in terms of power consumption and transaction speeds/fees. In addition, regulatory authorities at the regional, national, and international levels need to develop new and updated versions of regulatory frameworks to provide a better support for the next generation energy domain DLT applications. In addition, standardization is still needed to simplify integrating DLT applications and blockchain platforms from different vendors.

References

1. Bahrami, S., & Amini, M. H. (2017). A decentralized framework for real-time energy trading in distribution networks with load and generation uncertainty. arXiv preprint arXiv:1705.02575.
2. World Energy Council. (2018) World Energy Insights Brief 2018, Is Blockchain in Energy Driving an Evolution or a Revolution? London.
3. Discussion Paper: Applying Blockchain Technology to Electric Power Sector, Council on Foreign Relations, Accessed on: Available: https://www.cfr.org/report/applying-blockchain-technology-electric-power-systems
4. P. Richard, S. Mamel, L. Vogel, J. Strueker & L. Einhellig, "DENA Multi-stakeholder Study: Blockchain in the Integrated Energy Transition", DENA, 2019, Accessed on: Available: https://www.dena.de/newsroom/publikationsdetailansicht/pub/blockchain-in-the-integrated-energy-transition/
5. Cali, U., & Lima, C. (2020). Energy informatics using the distributed ledger technology and advanced data analytics. In Cases on Green Energy and Sustainable Development (pp. 438–481). IGI Global.
6. Cali, U., Lima, C., Li, X., & Ogushi, Y. (2019, July). DLT/blockchain in transactive energy use cases segmentation and standardization framework. In 2019 IEEE PES Transactive Energy Systems Conference (TESC) (pp. 1–5). IEEE.
7. https://www.energyweb.org/
8. https://sustainability.google/progress/projects/24x7/
9. https://news.microsoft.com/climate
10. https://www.powerledger.io/
11. https://ekwateur.fr/
12. https://www.alperia.eu/dev
13. https://www.ledgerinsights.com/ptt-blockchain-renewable-energy-marketplace-rec-ewf/
14. https://www.spgroup.com.sg/wcm/connect/spgrp/8827533d-e023-43dc-b744-6ede80f90416/%5B20191031%5D+Media+Release+-+SP+Group+And+I-REC+Help+Corporates+Achieve+Green+Targets.pdf?MOD=AJPERES&CVID=
15. https://lo3energy.com/
16. https://www.powerledger.io/
17. https://www.greeneum.net/
18. https://verv.energy/research
19. https://quantoz.com/wp-content/uploads/2020/10/QUASAR-whitepaper.pdf
20. https://www.energy21.com/
21. MOBI, "Mobility Open Blockchain Initiative". [Online]. https://dlt.mobi/ (Accessed 2021-03-16).
22. Energy Web Foundation, "Share&Charge Foundation Launches Open Charging Network on Energy Web Chain". [Online]. https://medium.com/energy-web-insights/share-charge-foundation-launches-open-charging-network-on-energy-web-chain-84af7f6f5b3c (Accessed 2021-03-16).
23. Open Charging Network, "Open Charging Network Wiki". [Online]. https://shareandcharge.atlassian.net/wiki/spaces/OCN/overview (Accessed 2021-03-16).
24. N. Lasla, et. al., "Blockchain Based Trading Platform for Electric Vehicle Charging in Smart Cities". [Online]. https://ieeexplore.ieee.org/stamp/stamp.jsp?tp=&arnumber=9125997 (Accessed 2021-03-16).

25. Blockchain in Japan, https://www.eu-japan.eu/sites/default/files/publications/docs/blockchaininjapan-martagonzalez.pdf
26. Drift Is a New Startup Applying Peer-to-Peer Trading to Retail Electricity Markets, https://www.greentechmedia.com/articles/read/drift-is-a-startup-applying-peer-to-peer-trading-to-retail-electricity
27. Quantoz and Energy21 https://quantoz.com/blog/quantoz-and-energy21-jointly-develop-concept-for-the-layered-energy-system-and-start-pilot-project-with-grid-operator-stedin/
28. https://eventhorizonsummit.com/data/uploads/2019/04/2019-Energy-Blockchain-Startups-Who-is-Who-.pdf
29. Sunchain. Available online: http://www.Sunchain.fr (Accessed 2019-03-16)
30. https://www.nrcan.gc.ca/science-and-data/funding-partnerships/funding-opportunities/current-investments/transactive-grid-enabling-end-end-market-services-framework-using-blockchain/22137
31. https://tokenpost.com/Interac-Alectra-Utilities-to-incentivize-people-to-use-renewable-energy-using-blockchain-2439
32. [http://bankymoon.co.za/]
33. Dutch energy supplier BAS to accept Bitcoin https://www.ccn.com/dutch-energy-supplier-bas-to-accept-bitcoin/
34. Dutch energy supplier BAS Nederland announces Bitcoin as its newest payment option https://99bitcoins.com/dutch-energy-supplier-bas-nederland-bitcoin-payment-option
35. Blockchain in the Mexican Energy Sector https://www.energypartnership.mx/fileadmin/user_upload/mexico/media_elements/reports/Blockchain_in_the_Mexican_Energy_Sector.pdf
36. German regional utility Enercity accepts Bitcoin https://blog.energybrainpool.com/en/german-regional-utility-enercity-accepts-bitcoin/
37. LEDGER, (https://www.ledger.com/)
38. ENGIE and Ledger, https://gems.engie.com/business-news/engie-and-ledger-to-harness-blockchain-technology-to-connect-the-physical-energy-and-digital-worlds/
39. M. Pilkington, "Blockchain technology: principles and applications," Research Handbook on Digital Transformations, pp. 1–39, 2015
40. H. Sukhwani, N. Wang, K. S. Trivedi and A. Rindos, "Performance Modeling of Hyperledger Fabric (Permissioned Blockchain Network)," In Proc. 2018 IEEE 17th International Symposium on Network Computing and Applications (NCA), Cambridge, MA, 2018, pp. 1–8.
41. Q. Nasir, I. A. Qasse, M. A. Talib, and A. B. Nassif, "Performance Analysis of Hyperledger Fabric Platforms," Security and Communication Networks, vol. 2018, Article ID 3976093, 14 pages, 2018. https://doi.org/10.1155/2018/3976093.
42. P. Thakkar, S. Nathan and B. Viswanathan, "Performance Benchmarking and Optimizing Hyperledger Fabric Blockchain Platform," 2018 IEEE 26th International Symposium on Modeling, Analysis, and Simulation of Computer and Telecommunication Systems (MASCOTS), Milwaukee, WI, 2018, pp. 264–276.
43. E. Androulaki, A. Barger, V. Bortnikov, C. Cachin, K. Christidis, A. D. Caro, D. Enyeart, C. Ferris, G. Laventman, Y. Manevich, S. Muralidharan, C. Murthy, B. Nguyen, M. Sethi, G. Singh, K. Smith, A. Sorniotti, C. Stathakopoulou, M. Vukolic, S. W. Cacco and J. Yellick, "Hyperledger Fabric: A Distributed Operating System for Permissioned Blockchains," In Proc. 2018 the Thirteenth EuroSys Conference (EuroSys'18), No. 30, Porto, Portugal, April 23–26, 2018.
44. Hyperledger Performance and Scale Working Group, Hyperledger Blockchain Performance Metrics, 2018 [Online]. Available: https://www.hyperledger.org/wp-content/uploads/2018/10/HL_Whitepaper_Metrics_PDF_V1.01.pdf. Retrieved: April 2019.
45. S. Rouhani and R. Deters, "Performance analysis of ethereum transactions in private blockchain," In Proc. 2017 8th IEEE International Conference on Software Engineering and Service Science (ICSESS), Beijing, China, 2017, pp. 70–74.
46. S. Chen, J. Zhang, R. Shi, J. Yan and Q. Ke, "A Comparative Testing on Performance of Blockchain and Relational Database: Foundation for Applying Smart Technology into Current

Business Systems," In Proc. 2018 Int'l Conference on Distributed, Ambient and Pervasive Interactions, pp. 21–34, Las Vegas, July 15–20, 2018.

47. A. Roehrs, C. A. Costa, R. R. Righi, V. F. Silva, J. R. Goldim and D. C. Schmidt, "Analyzing the performance of a blockchain-based personal health record implementation," Journal of Biomedical Informatics, Vol. 92, April 2019.

48. Kuzlu, M., Pipattanasomporn, M., Gurses, L., & Rahman, S. (2019, July). Performance analysis of a hyperledger fabric blockchain framework: throughput, latency and scalability. In 2019 IEEE international conference on blockchain (Blockchain) (pp. 536–540). IEEE.

Chapter 10
Optimization and Digitalization of Power Markets

Electricity is essential and fundamental to society and people. The power industry over the globe has been undergoing major transitions toward deregulated competitive markets, which bring more affordable, reliable, and cleaner electric power to customers. The wholesale and retail electricity markets have been established in many countries to allow power to be bought and sold, similar to other commodities [1]. Optimization and digitalization are the key technologies and enablers for power markets to realize energy and ancillary service trading and settlement.

In this chapter, we discuss the basics of power markets. Section 10.1 gives an introduction of the history of power markets and basic principles of market operations, including energy, capacity, and ancillary service markets. Section 10.2 introduces the mathematical modeling for market clearing and the concept of locational marginal pricing (LMP). In Section 10.3, we discuss the emerging distribution and local electricity markets. Also, the optimization model for distribution market clearing and distributional locational marginal pricing (DLMPs) is discussed. Section 10.4 summarizes this chapter.

10.1 Introduction of Power Markets

Since the 1990s, many countries have started to reform the power industry from regulated, to deregulated competitive wholesale electricity markets. Before that, most investor-owned utilities were vertically integrated and regulated, including both generation and transmission. In the wholesale power markets at that time, bilateral trading was the main form of electricity trading. It is common that some utilities purchase excessive electric power from others rather than generate power by their own generators [2]. In the US, the wholesale market and trading activities are monitored and regulated by the Federal Energy Regulatory Commission (FERC). Since the early 1990s, the US started restructuring the power grid to

© The Author(s), under exclusive license to Springer Nature Switzerland AG 2021
U. Cali et al., *Digitalization of Power Markets and Systems Using Energy Informatics*, https://doi.org/10.1007/978-3-030-83301-5_10

deregulate the power industry and create competitive power markets for more affordable and reliable clean power. In 1996, FERC issued Orders No. 888 [3] and No. 889 [4], which accelerated the process of deregulation and open generation to competition. Until now, several regional wholesale power markets have been established in the US and operated by regional transmission operators (RTOs) or independent system operators (ISOs), including California ISO, Southern Power Pool (SPP), Electric Reliability Council of Texas (ERCOT), Midcontinent ISO (MISO), PJM Interconnection, New York ISO, and ISO New England as shown in Fig. 10.1. In addition, Canada has two ISOs, including Alberta Electric System Operator (AESO) and Independent Electricity System Operator (IESO). Other countries that have established electricity markets include Australia, New Zealand, UK, Norway, etc.

A RTO or ISO is a nonprofit and independent organization that has three critical roles in grid operation, market administration, and system planning. The ISO is responsible for the optimal scheduling and dispatch of the generation resources and transmission network for reliable grid operation, market clearing, pricing and settlement, and system planning for resource adequacy. In the established centralized wholesale markets, ISOs/RTOs operate three types of markets including:

Fig. 10.1 Map of North America ISO markets [5]

- Energy markets for day-to-day wholesale electric power buying and selling among the market participants. The power producers offer to sell power to load-serving entities, who then resell it to consumers.
- Ancillary service market for trading grid services that ensure short-term system reliability.
- Capacity market for ensuring long-term generation resource adequacy and system reliability.

Through deregulation, the competition in the power market has brought numerous benefits, including [6]

- Lower-power production cost. The competition in the market will drive power producers to reduce their generation costs. The customers will have access to lower-cost electric power.
- Grid investment decisions are made based on market prices. The price signal provided by the power market gives incentive to the investors for decision-making in the types and locations of the resources. This practice will drive the investors to invest in generation/transmission assets at locations that are beneficial to the grid operation in the long term. Also, it will maximize the revenue and improve the return on investment.
- Reliability enhancement. The ISO/RTO can coordinate the resources owned by different entities to manage the system and respond to system incidents and disturbances to improve system reliability.

Retail Market: The utilities, load-serving entities (LSEs), or retailers can buy electric power from the wholesale market and then resell it to end users and consumers in the retail market. A simple example of the power purchase and sale in the retail market is that a household pays for its electricity usage to a local power utility month to month.

The retailers play as market participants in both wholesale market and retail markets. In the wholesale market, it is a buyer that purchases electricity from power producers. In the retail market, it is a seller that sells electricity to the end-consumers at a flat rate, time-of-use (TOU), or other tariffs. There are numerous options for consumers to purchase electricity, such as competitive retailers and local utilities as service providers.

The wholesale market costs are reflected on the electricity bills of end-customers as kWh charge, called "Basic Service" [7]. On top of that, customers also pay the delivery charge to the local utility company. This includes transmission, distribution, transition, and other charges approved by the state, who regulates all the utility companies. Retail markets have different customer categories, including residential, commercial, and industrial. The pricing for different classes of customers may be different. The utility rate structures and the relationship between the retail rate and wholesale market costs may vary according to state retail procurement policies.

10.1.1 Energy Market

The energy market coordinates the trading and production of electricity where the power producers offer to sell the electric power at a bid price, and the load-serving entities bid for the power that is needed for its own customers.

The wholesale market includes two energy markets in two different time scales, that is, the day-ahead market (DAM) and the real-time market (RTM). This is also called a "multi-settlement" system. The energy and ancillary services are usually co-optimized and jointly cleared in the electricity markets. Market participants can participate in multiple markets.

- The day-ahead energy market is a forward market in which market participants commit to buy or sell electricity one day before the operating day. The majority of market transactions happen in the day-ahead market based on the forecasted load. The ISO will clear the market based on the bids of power producers and determine the optimal dispatch of the generators to meet the load demand. The day-ahead market prices – locational marginal prices (LMPs) – will be posted by the ISO.
- The real-time energy market, also known as the balancing market, is used for the remaining transactions that take place to balance the actual demand and the day-ahead commitments during the operating day. The ISO will make real-time optimal dispatch decisions and post the real-time LMPs based on the market clearing of the real-time market.

10.1.2 Ancillary Service Market

Ancillary services are needed by the system operator to maintain a secure and reliable system operation and balance the system. Ancillary services typically include reserve and regulation that help maintain grid frequency and provide backup power. The ancillary service market is designed and operated by the ISO to procure regulation and reserve from generators and reward these services to the grid. Other ancillary services may include reactive power compensation and black start.

Regulation is used to balance very short-term (second-level) mismatches between generation and load to maintain real-time system balance and frequency. It is designed to handle fast load fluctuations and small generation variations. The ISO will reward market participants for offering regulation capacity that allows upward and downward headroom to increase or decrease power output in seconds to keep the real-time system balance. Regulation service can be provided by generating units that can quickly adjust its power output. The ISO will dispatch the resources of market participants for Automatic Generation Control (AGC) in order to maintain the area control error (ACE) and frequency of the system within acceptable bounds. In North America, regulation, as a market product, is used to maintain system frequency of 60 Hz, respond to balance second-to-second load demand variations.

Reserve services are designed to address large power imbalances caused by system disturbances and contingencies. Reserves represent the generation capacity that can be quickly available to dispatch for addressing ramping and unexpected system contingencies, such as the generator trip. Different ISOs may define different types of reserve products in their markets. Below are a few common types of reserves in the PJM market [8]:

Operating Reserve The amount of power that should be delivered within 30 min when called by the ISO. Operating reserve can be provided by

- Generating units that are connected to the grid or offline
- Demand response that can reduce the load

Primary Reserve The amount of power that should be delivered within 10 min when called by the ISO. It can be provided by

- Generating units that are connected to the grid or offline
- Demand response that can reduce the load

Synchronized Reserve Also known as Spinning Reserve, is the amount of power that should be delivered within 10 min when called by the ISO. It can be provided by

- Generating units that are connected to the grid
- Demand response that can reduce the load

Supplemental Reserve Also known as non-spinning reserve, is the amount of power that should be delivered within 10–30 min when called by the ISO. It can be provided by

- Generating units that are connected to the grid or offline
- Demand response that can reduce the load

The ancillary services are important to maintain the reliability of real-time system operation. A generator can bid in both energy and ancillary services simultaneously. In some electricity markets, the dispatch of energy and ancillary services is jointly optimized, and their transactions are jointly cleared by the ISO.

10.1.3   Capacity Market

Capacity markets are used in some ISOs, not all, to ensure resource adequacy and system reliability to meet the future load demand. Those ISOs include PJM, New York ISO, and ISO New England. Other ISOs, such as ERCOT, do not have a capacity market. The capacity market is designed to ensure long-term grid reliability in which the ISO determines the capacity of generation resources required to meet its reliability criteria or target in the next few years, for example, 3 years in the PJM market [9]. The power producers will bid to provide the capacity, and consumers must pay for their share of the capacity, and the price depends on the amount of offered capacity. The capacity market provides long-term capacity incentives and price signals to attract investment to ensure resource adequacy and maintain system reliability.

A few essential elements in the capacity market include

- The required capacity is procured through a competitive auction years before it is needed.
- The capacity prices vary from location to location (locational pricing) as they reflect the transmission constraints, and some locations are preferred by the system and ISO to have new capacity added.
- The resource requirement can be a variable curve, known as the variable resource requirement (VRR) curve in the PJM market, which the ISO used to procure a certain amount of capacity at each price point.

The capacity market mechanisms or rules may vary in different ISOs and markets. In addition to new generation capacity, the resources that can be offered by the market participant may include demand response, energy efficiency, etc. They must commit to increase the generation capacity or reduce the load demand by the amount they offered in the capacity market.

10.2   Wholesale Market Clearing and Pricing

In the wholesale market, the power producers submit offers, and load servicing entities (LSEs) or retailers will submit bids to the ISO. Most of the time, the load demand is inelastic, so in the wholesale market, the ISO will mainly decide to dispatch which generators to meet the total demand. In the day-ahead market, ISOs receive the bids from power producers and determine the optimal dispatch of the generators for the next day in an order from lowest to the highest marginal generation cost. For example, some renewable-based power generators do not have a fuel cost, and they can bid at $0/MWh into the wholesale energy market, and the ISO will dispatch them first as they are the least expensive ones. This will allow the market to meet the total load demand by the lowest system cost and offer affordable electricity to consumers [10]. During heavy load periods, more expensive generators need to be started up or dispatched, which will raise the electricity prices overall. At locations where the transmission lines are congested, the electricity prices will be higher than other places. Therefore, electricity market prices are locational-temporal differentiated depending on the operation conditions of the transmission system.

The market clearing is the process of the ISOs to clear the bids and determine the optimal dispatch of generators in the market, and accordingly, the energy prices will be calculated based on the marginal cost principle. That is, the nodal energy price, referred to as locational marginal price (LMP), is calculated based on the marginal cost it brings to the system for serving the next MW load at a specific location. The optimal dispatch of the generators is based on the bids submitted by the generators, including the bidding price and quantity. The bidding price of a generator is assumed to be its marginal generation cost. With the bidding information and the load forecasts, the ISO performs security-constrained unit commitment (SCUC) in the day-ahead market to determine which generators and when they will be on and off. Then security-constrained economic dispatch (SCED) is used to determine the optimal dispatch (the amount of power generation) of each unit. Based on the optimal solution, the nodal market prices can be calculated. All market participants at this location will buy/sell electricity at the energy prices determined by the ISO. After the market clearing, the ISO will issue the schedules to the generators and post the energy prices. The procedure of the ISO market operation is shown in Fig. 10.2.

Basically, wholesale market prices typically reflect the price for power when it is able to flow freely without transmission constraints across the ISO's territory.

10.2.1   Optimal Power Flow and Economic Dispatch

The direct-current optimal power flow (DCOPF) and economic dispatch models are the keys of the ISO market clearing process. It will determine the optimal scheduling of each generator, considering the network security constraints.

The objective of the power market is to maximize social welfare. The electric power load is mostly inelastic, leading the objective function to be equivalent to

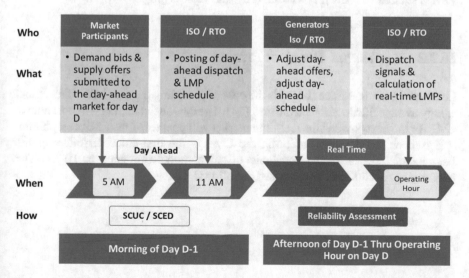

Fig. 10.2 ISO market operation framework. Ellison, J. F., Tesfatsion, L., Loose, V. W., Byrne, R. H. (2012). Project Report: A Survey of Operating Reserve Markets in U.S. ISO/RTO-Managed Electric Energy Regions, Sandia Report, SAND2012-1000, unlimited release, September

minimizing the total generation cost. The control variables are the generator dispatch. The constraints include the energy balance equation, generation limits, and transmission power flow limits, as shown in the mathematical model below [10].

$$\min \quad \sum_{i=1}^{N} c_i \cdot G_i$$

$$\text{s.t.} \quad \sum_{i=1}^{N} G_i = \sum_{i=1}^{N} D_i$$

$$\sum_{i=1}^{N} \text{GSF}_{k-i} \cdot \left(G_i - D_i\right) \leq \text{PLimit}_k, \quad k = 1, 2, \ldots, M$$

$$G_i^{\min} \leq G_i \leq G_i^{\max}, \quad i = 1, 2, \ldots, N$$

(10.1)

where

N: number of buses
M: number of transmission lines
c_i: generation (bid) cost of generator at Bus i ($/MWh)
G_i: generation dispatch at Bus i (MWh)
$G_i^{\max}$: maximum generation output at Bus i
$G_i^{\min}$: minimum generation output at Bus i
D_i: demand at Bus i (MWh)
GSF_{k-i}: generation shift factor to line k from Bus i
PLimit_k: transmission limit of Line k

This DCOPF-based economic dispatch model is essentially a linear programming optimization model that can be easily solved by off-the-shelf commercial solvers such as CPLEX [11], Gurobi [12], and Xpress [13].

10.2.2 Locational Marginal Pricing

Locational Marginal Price (LMP) is the cost of providing the next MW of electrical energy at a specific location on the grid [14]. The day-ahead market will post hourly LMP for the next day, and the real-time market will post real-time LMP every 5 min.

Based on the definition of LMP, the LMP is the first-order derivative of the Lagrangian function of the economic dispatch model presented in 10.2.1. The Lagrangian function can be expressed as follows:

$$L = \sum_{i=1}^{N} c_i \cdot G_i - \lambda \left(\sum_{i=1}^{N} G_i - \sum_{i=1}^{N} D_i \right)$$

$$- \sum_{k=1}^{N} \mu_k \left(\sum_{i=1}^{N} \text{GSF}_{k-i} \cdot (G_i - D_i) - \text{PLimit}_k \right) \tag{10.2}$$

$$- \sum_i \xi_+ \left(G_i - G_i^{\max} \right) - \sum_i \xi_- \left(G_i^{\min} - G_i \right)$$

where λ, μ, and ξ are dual variables or Lagrangian multipliers associated with the constraints in the DCOPF model.

Based on the Lagrangian function, the LMP at each bus can be calculated by the first-order derivative of the Lagrangian function with respect to the load demand on the bus.

$$\text{LMP}_B = \frac{\partial L}{\partial D_B} = \lambda + \sum_{k=1} \mu_k \text{GSF}_{k-B} \tag{10.3}$$

In the LMP formulation, the first term represents the energy price, and the second term represents the congestion price associated with the transmission constraints. The network loss can be represented by the delivery factor and hence will introduce a loss component in the LMP. The details regarding a loss DCOPF based market-clearing model can be found in [10].

In the case that no transmission constraints are binding (no transmission lines are congested), the LMPs on different nodes across the whole network are the same, and there will be only one marginal unit in the system. At this time, the total customer payment is equal to the total generation revenue. When there are transmission congestions in the system, the LMPs at the end nodes of a congested line will be different, and there could be multiple marginal units in the system, which will directly determine the LMPs at the nodes where the marginal units are located. Due to the congestion, the total customer payment will be larger than the total generation revenue. The difference (Total Customer Payment – Total Generation Revenue) is called congestion surplus. The congestion surplus cannot be held by the ISO. The ISO has to design a way to distribute the congestion surplus to market participants. The Financial Transmission Rights (FTR) mechanism was introduced to mitigate the price volatility caused by transmission congestion and can be utilized to distribute the congestion surplus.

Financial Transmission Right It can be observed that in the LMP mechanism, congestion cost will occur due to the congestion price component in the LMP. A financial instrument, Financial Transmission Right (FTR), is introduced for market participants to hedge the risk of potential congestion cost and be entitled to receive compensation in the day-ahead market when there are transmission congestions and the LMPs at two end nodes in a transmission path are different. Forward FTR auction takes place annually and monthly in which the market participants can submit bids to purchase FTR. The holders of FTR will be paid by the ISO based on the MW amount of the FTR and the locational differences in hourly congestion prices.

Information and control technologies (ICT) and digitalization are the key enablers of the power market to manage energy trading and settlement and coordinate physical operation and financial transactions. Corresponding software systems, such as Energy Management Systems (EMS) and Market Management Systems (MMS), and communication systems need to be streamlined for the power market. This allows for the ISO to send dispatch signals to all the generators and LMPs to all the market participants and monitor the operation status of the system through the Supervisory control and data acquisition (SCADA) system and phasor measurement units (PMUs). With the evolvement of market design, the software and ICT components need to upgrade and advance to meet the business and operation requirements.

Future wholesale markets should be able to integrate more uncertain renewables in the transmission power grids. Stochastic optimization and robust optimization-based approaches have been applied to DCOPF for stochastic market clearing in some research work [15–17]. Risk-based market mechanisms [18–19] and more accurate solar/wind forecasting will be needed to mitigate the risk in future power grids with high penetration of renewables.

10.3 Distribution and Local Energy Markets

10.3.1 The Emerging Distribution Electricity Markets

The fast-growing distributed energy resources (DERs) are driving the evolution of smart distribution power grids [20]. Various types of DERs, including distributed PVs and wind, distributed energy storage, electric vehicles, smart buildings, and demand response, are being integrated into the distribution power grid. The DERs are turning traditional consumers into prosumers and traditional passive distribution power grids into active distribution networks. With increasing numbers of DERs being integrated into the distribution power grid, they may cause potential voltage and network congestion issues. To efficiently manage enormous DERs and prosumers, new business models are needed to provide effective price signals to incentivize DERs to operate in a way that is beneficial to the grid.

Distribution and local electricity markets are emerging as a new paradigm to integrate, manage, and utilize large amounts of DERs in distribution networks [21, 22]. Establishing a distribution electricity market and introducing explicit market mechanisms to distribution systems will create a platform that allows transparent, efficient, and fair energy transactions among various market participants [23], that is, prosumers and DERs. In a distribution market, a distribution system operator (DSO) receives the bids from all market participants, clears the market, and makes financial settlements. The distribution-level market will unlock the value of the DERs as grid assets to the distribution power grid. In the long run, a well-designed distribution market and price signals could motivate rational investment and deployment of future DERs [24].

The distribution-level electricity market and distribution locational marginal price (DLMP) have been investigated in some research work. In [25], a day-ahead distribution market-clearing model was proposed, based on which DLMPs are calculated and decomposed into five components, including marginal costs for energy, reactive power, voltage support, congestion, and loss. It was found that, under the DLMP mechanism, DERs can contribute to voltage support and congestion management in distribution networks. [26] presented a reliability pricing mechanism in distribution networks that can identify and recover long-run investment costs. [27–29] developed DLMPs to address network congestion in distribution networks with distributed generators (DGs). DLMP was developed in [30, 31] to manage congestion caused by charging of high penetration EVs in future distribution power grids. Similarly, a DLMP-based congestion price was proposed to manage household demand response in [32]. Some of these existing methods for network congestion management are based on the DCOPF model without considering reactive power and voltage issues. [33] presented a co-optimization model for power and reserve in dynamic transmission and distribution markets considering renewable power generation and DERs. To allocate the cost for loss and benefit of emission reduction among the DERs in distribution networks, an LMP-based method was discussed in [34, 35].

Market-oriented reforms and practical implementations for distribution systems have been initiated in a number of countries. In the US, New York State issued the "Reforming the Energy Vision" initiative [36], which includes a distributed system platform (DSP) for integrating high-penetration DERs. California has initiated multiple proceedings for the development of "distributed resource plans" [37]. Grid modernization plans have been implemented in the utilities in many other states such as Hawaii, Minnesota, and Massachusetts [38].

10.3.2   Framework of Distribution-Level Electricity Markets

The framework of a distribution-level electricity market operated by the distribution system operator (DSO) is shown in Fig. 10.3. In the distribution power grid of the future, various types of DERs owned by different entities may participate in the distribution electricity market. They can participate in the distribution market individually or in the forms of aggregators, demand response, microgrids, and virtual power plants (VPPs). They can submit bids to the DSO. The DSO will perform distribution market clearing to clear the bids of all market participants meanwhile considering the interactions with the wholesale market. Essentially, the DSO is also a market participant that trades energy in the wholesale market. Depending on the penetrations of DERs and time of the day, it can either buy energy from or sell energy to the wholesale market. In addition to energy trading, the distribution market has to take reactive power into account as a market product as it is important to

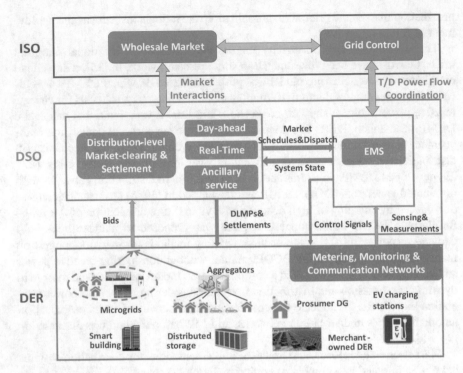

Fig. 10.3 The framework of the distribution-level electricity market [26]

maintain the voltage of a distribution network. Thus, the DSO will joint clear the active power (energy) and reactive power in the distribution market and post-DLMPs to manage the operation of DERs.

In addition, the voltage control devices and network topology have a significant impact on the bidirectional power flow in the distribution power grid with DERs and thus will impact the optimal dispatch and market clearing in distribution systems. Voltage control technologies such as on-load tap changer (OLTC), shunt capacitor banks (CB), and network reconfiguration shall be incorporated into the DSO optimization and market clearing.

10.3.3   Market Clearing and Pricing in Distribution-Level Electricity Markets

Based on the distribution market framework, the day-ahead market-clearing model for the distribution market can be formulated based on the distribution network optimal flow model. Typical AC optimal power flow models used in distribution networks include the Distflow model and its linearized form [39] and second-order-cone-based convex model [40]. The objective is to maximize the social welfare or minimize the total operation cost considering the voltage and thermal security constraints of the distribution network.

The market-clearing process in the day-ahead distribution market may include two steps. In the first stage, DSO runs a market scheduling model to determine the optimal settings for OLTC [41], network configuration [42], and capacitor banks. In the second stage, DSO will run the market pricing model to optimally dispatch the DERs and calculate the DLMPs for both active power and reactive power. Ancillary services can also be defined and incorporated. This procedure can account for the features and controls in distribution networks into the market framework.

Based on the Lagrangian function of the market pricing model, the DLMP can be calculated and explicitly decomposed into five components, including energy, reactive power, congestion, voltage support, and loss prices. With these elements contained in the DLMP, voltage and congestion management can be achieved with DERs responding to the price signals.

The design and mechanisms of distribution markets and DLMPs still remain an open question. To manage and facilitate the distribution market operation, advanced ICT and digitalization technologies and solutions are key enablers to incorporate large amounts of DERs into distribution system operations and markets.

10.3.4 Local Energy Market and Peer-to-Peer Energy Trading

With the increasing penetration of large-scale small amounts of DERs into the electric power grid, the traditional centralized power system operation is being transformed into a decentralized, bottom-up, and localized control paradigm. With more DERs, the local energy market is one of the solutions to integrate DERs by providing effective pricing and incentive mechanisms.

With the increasing deployment of DERs, traditional consumers are becoming prosumers. Peer-to-peer (P2P) technology has been utilized for local energy trading in several projects. In the Piclo project in the UK, the business consumers can buy electricity from local renewable power producers and the customers can select and prioritize the generators that they will purchase power from. It provides contracts, meter data, billing, and balances the market. The P2P technology mainly relies on the DLT technology such as blockchain that has been introduced and discussed in our previous chapters. In [43], various designs of blockchain technology for P2P energy trading were evaluated in aspects of privacy, security, market acceptance, and so on. The study in [44] was focused on evaluating the security and performance of a blockchain-based P2P platform for energy trading under various conditions. Specifically, in [45], a blockchain-based platform was proposed and designed for a local energy market, which has 100 residential prosumers and consumers. In this design, there is no need for a central mediator or intermediary to facilitate the power trading and exchange. Decentralized optimization approaches such as the alternating direction method of multipliers (ADMM) for supporting the P2P energy trading have been investigated in [46, 47].

10.3.5 Market Integrations of DERs

The FERC Order No. 2222 was issued in 2020 to allow DERs and DER aggregators to participate in the wholesale energy, ancillary service, and capacity markets operated by ISOs/RTOs. The DERs and DER aggregators can participate in the ISO markets if the aggregation is at least 100 kW in size and the aggregation can be geographically broad as technically feasible [48]. The data, bidding, metering, and telemetry for DER aggregators need to be aligned with the existing requirement. In the future hierarchical structure of electricity markets, including wholesale, distribution, and local electricity markets, the coordination between the markets at different levels needs to be carefully addressed.

There are several key elements that need to be considered in the coordination.

- The coordination between DER aggregators and distribution utilities.
- Distribution utility may override the ISO/RTO dispatch/schedule of DER aggregators to ensure the safety and reliability of distribution systems.
- Data sharing and communication requirements and practices.

10.4 Summary

In this chapter, the optimization and digitalization of the power market were reviewed and introduced. Starting from the established wholesale market, the energy, ancillary service, and capacity markets are reviewed. The DCOPF-based economic dispatch and market-clearing model and the resulting LMP formulation are introduced. Further, the emerging distribution and local electricity markets are reviewed. The technical challenges and new technologies such as peer-to-peer and DLT technology are presented. The future electricity market may involve multiple levels of markets and participants, which is also discussed in the chapter.

References

1. www.pjm.com.
2. US Electricity Markets 101. https://www.rff.org/publications/explainers/us-electricity-markets-101/
3. "FERC: Landmark Orders - Order No. 888". www.ferc.gov. Retrieved 14 May 2017.
4. "Order No. 889, Final Rule, FERC.gov". Retrieved 14 May 2017.
5. https://en.wikipedia.org/wiki/Regional_transmission_organization_(North_America)
6. Judy Chang, Johannes Pfeifenberger, John Tsoukalis, "Potential Benefits of a Regional Wholesale Power Market to North Carolina's Electricity Customers", Brattle Group, 2019. https://brattlefiles.blob.core.windows.net/files/16092_nc_wholesale_power_market_whitepaper_april_2019_final.pdf.
7. https://www.iso-ne.com/about/what-we-do/in-depth/wholesale-vs-retail-electricity-costs

8. https://learn.pjm.com/three-priorities/buying-and-selling-energy/ancillary-services-market/reserves.aspx

9. https://learn.pjm.com/three-priorities/buying-and-selling-energy/capacity-markets.aspx#:~:text=PJM's%20capacity%20market%2C%20called%20the,three%20years%20in%20the%20future.

10. F. Li and R. Bo, "DCOPF-based LMP simulation: Algorithm, comparison with ACOPF, sensitivity," IEEE Trans. Power Syst., vol. 22, no. 4, pp. 1475–1485, Nov. 2007.

11. Blieklú C, Bonami P, Lodi A. Solving mixed-integer quadratic programming problems with IBM-CPLEX: a progress report. InProceedings of the twenty-sixth RAMP symposium 2014 Oct (pp. 16–17).

12. Anand R, Aggarwal D, Kumar V. A comparative analysis of optimization solvers. Journal of Statistics and Management Systems. 2017 Jul 4;20(4):623–35.

13. Berthold T, Farmer J, Heinz S, Perregaard M. Parallelization of the FICO xpress-optimizer. Optimization Methods and Software. 2018 May 4;33(3):518–29.

14. https://www.misoenergy.org/stakeholder-engagement/training2/learning-center/market-basics/

15. Bouffard F, Galiana FD, Conejo AJ. Market-clearing with stochastic security-part I: formulation. IEEE Transactions on Power Systems. 2005 Oct 31;20(4):1818–26.

16. Zavala VM, Kim K, Anitescu M, Birge J. A stochastic electricity market clearing formulation with consistent pricing properties. Operations Research. 2017 Jun;65(3):557–76.

17. Lei M, Zhang J, Dong X, Jane JY. Modeling the bids of wind power producers in the day-ahead market with stochastic market clearing. Sustainable Energy Technologies and Assessments. 2016 Aug 1;16:151–61.

18. Zhang N, Kang C, Xia Q, Ding Y, Huang Y, Sun R, Huang J, Bai J. A convex model of risk-based unit commitment for day-ahead market clearing considering wind power uncertainty. IEEE Transactions on Power Systems. 2014 Oct 8;30(3):1582–92.

19. Liu M, Wu FF. Risk management in a competitive electricity market. International Journal of Electrical Power & Energy Systems. 2007 Nov 1;29(9):690–7.

20. G. T. Heydt, B. H. Chowdhury, M. L. Crow, D. Haughton, B. D. Kiefer, F. Meng, and B. Sathyanarayana, "Pricing and control in the next generation power distribution system," IEEE Trans. Smart Grid, vol. 3, no. 2, pp. 907–914, June 2012.

21. S. Bahramirad, A. Khodaei, and R. Masiello, "Distribution markets", IEEE Power & Energy Magazine, vol. 14, no.2, pp.102–106, 2016.

22. F. Rahimi and S. Mokhtari, "From ISO to DSO: imagining new construct--an independent system operator for the distribution network," Public Util. Fortn., vol. 152, no. 6, pp. 42–50, 2014.

23. E. Ela, J. Riesz, J. O'Sullivan, B. F. Hobbs, M. O'Malley, M. Milligan, P. Sotkiewicz, J. Caldwell, "The evolution of the market: designing a market for high levels of variable generation", IEEE Power & Energy Magazine, vol. 13, no. 6, pp. 60–66, 2015.

24. Parhizi S, Khodaei A, Bahramirad S. Distribution market clearing and settlement. In2016 IEEE Power and Energy Society General Meeting (PESGM) 2016 Jul 17 (pp. 1–5). IEEE.

25. Bai L, Wang J, Wang C, Chen C, Li F. Distribution locational marginal pricing (DLMP) for congestion management and voltage support. IEEE Transactions on Power Systems. 2017 Oct 30;33(4):4061–73.

26. C. Gu, J. Wu and F. Li, "Reliability-based distribution network pricing", IEEE Trans. Power Syst., vol. 27, no. 3, pp. 1646–1655, 2012.

27. P. M. Sotkiewicz and J. M. Vignolo, "Nodal pricing for distribution networks: efficient pricing for efficiency enhancing DG," IEEE Trans. Power Syst., vol. 21, no. 2, pp. 1013–1014, May 2006.

28. F. Meng and B. H. Chowdhury, "Distribution LMP-based economic operation for future smart grid," in Proc. 2011 IEEE Power and Energy Conference at Illinois, pp. 1–5, 2011.

29. S. Huang, Q. Wu, S. S. Oren, R. Li, and Z. Liu, "Distribution locational marginal pricing through quadratic programming for congestion management in distribution networks," IEEE Trans. Power Syst., vol. 30, no. 4, pp. 2170–2178, Jul. 2015.

30. R. Li, Q. Wu, and S. S. Oren, "Distribution locational marginal pricing for optimal electric vehicle charging management," IEEE Trans. Power Syst., vol. 29, no. 1, pp. 203–211, Jan. 2014.

31. W. Liu, Q. Wu, F. Wen, and J. Østergaard, "Day-ahead congestion management in distribution systems through household demand response and distribution congestion prices," IEEE Trans. Smart Grid, vol. 5, no. 6, pp. 2739–2747, Jul. 2014.

32. J. Momoh, X. Yan, and G. D. Boswell. "Locational marginal pricing for real and reactive power." IEEE Power and Energy Society General Meeting, Pittsburgh, PA, pp. 1–6, 2008.

33. M. Caramanis, E. Ntakou, W. Hogan, A. Chakrabortty, and J. Schoene, "Co-optimization of power and reserves in dynamic T&D power markets with nondispatchable renewable generation and distributed energy resources", Proceedings of the IEEE, vol. 104, no. 4, pp. 807–836, 2016.

34. K. Shaloudegi, N. Madinehi, S. H. Hosseinian, and H. A. Abyaneh, "A novel policy for locational marginal price calculation in distribution systems based on loss reduction allocation using game theory," IEEE Trans. Power Syst., vol. 27, no. 2, pp. 811–820, May 2012.

35. E. Farsani, H. Abynaeh, M. Abedi, and S. H. Hosseinian, "A novel policy for LMP calculation in distribution networks based on loss and emission reduction allocation using nucleolus theory", IEEE Trans. Power Syst., vol. 31, no.1, pp. 163–152, 2016.

36. New York State, "Reforming the energy vision," NYS Dept. Public Service, Albany, NY, USA, 2014.

37. California State, "Distribution resources plan", California Public Utility Commission, CA, USA, 2014.

38. Minnesota: Minnesota Department of Commerce, Value of Solar Tariff Methodology, Dec. 2014.

39. Gilbert GM, Bouchard DE, Chikhani AY. A comparison of load flow analysis using DistFlow, Gauss-Seidel, and optimal load flow algorithms. InConference Proceedings. IEEE Canadian Conference on Electrical and Computer Engineering (Cat. No. 98TH8341) 1998 May 25 (Vol. 2, pp. 850–853). IEEE.

40. M. Farivar and S.H. Low, "Branch flow model: relaxations and convexification–parts I", IEEE Trans. Power Syst., vol. 28, no. 3, pp. 2554–2564, 2013.

41. Z. Wang, H. Chen. J. Wang, M. Begovic, "Inverter-less hybrid voltage/var control for distribution circuits with photovoltaic generators", IEEE Trans. Smart Grid, vol. 5, no. 6, pp. 2718 – 2728, 2014.

42. R. Jabr, R. Singh and B. Pal, "Minimum loss network reconfiguration using mixed-integer convex programming," IEEE Trans. Power Syst., vol. 27, no. 2, pp. 1106–1115, 2012.

43. Vangulick D, Corn'elusse B, Ernst D. Blockchain for peer-to-peer energy exchanges: design and recommendations. Power Syst Comput Conf (PSCC) 2018;2018:1–7.

44. Aitzhan NZ, Svetinovic D. Security and privacy in decentralized energy trading through multi-signatures, blockchain and anonymous messaging streams. IEEE Trans Dependable Secure Comput 2016;15:840–52.

45. Mengelkamp E, Notheisen B, Beer C, Dauer D, Weinhardt C. A blockchain-based smart grid: towards sustainable local energy markets. Comput Sci-Res Dev 2018; 33:207–14.

46. K. Zhang, S. Troitzsch, S. Hanif, and T. Hamacher. Coordinated Market Design for Peer-to-Peer Energy Trade and Ancillary Services in Distribution Grids. IEEE Transactions on Smart Grid. (Early Access). DOI: https://doi.org/10.1109/TSG.2020.2966216.

47. D. Nguyen, "Optimal solution analysis and decentralized mechanisms for peer-to-peer energy markets," IEEE Transactions on Power Systems, in-press, 2020.

48. "Participation of Distributed Energy Resource Aggregations in Markets Operated by Regional Transmission Organizations and Independent System Operators", FERC, 2020. https://www.federalregister.gov/d/2020-20973

Index

Printed in the United States
by Baker & Taylor Publisher Services